Basic Structures

Basic Structures

2nd Edition

Philip Garrison

BSc, MBA, CEng, MICE, MIStructE, MCIHT
School of the Built Environment
Leeds Metropolitan University

WILEY-BLACKWELL

A John Wiley & Sons, Ltd., Publication

This edition first published 2011
© 2005 Philip Garrison
© 2011 by John Wiley & Sons, Ltd

Wiley-Blackwell is an imprint of John Wiley & Sons, formed by the merger of Wiley's global Scientific, Technical and Medical business with Blackwell Publishing.

First edition published 2005
Second edition published 2011

Registered Office
John Wiley & Sons, Ltd, The Atrium, Southern Gate, Chichester, West Sussex, PO19 8SQ, UK

Editorial Offices
9600 Garsington Road, Oxford, OX4 2DQ, UK
The Atrium, Southern Gate, Chichester, West Sussex, PO19 8SQ, UK
2121 State Avenue, Ames, Iowa 50014-8300, USA

For details of our global editorial offices, for customer services and for information about how to apply for permission to reuse the copyright material in this book please see our website at www.wiley.com/wiley-blackwell.

The right of the author to be identified as the author of this work has been asserted in accordance with the UK Copyright, Designs and Patents Act 1988.

Library of Congress Cataloging-in-Publication Data

Garrison, Philip, CEng.
Basic structures / Philip Garrison. – 2nd ed.
    p.   cm.
  Revision and expansion of: Basic structures for engineers and architects. 2005.
  Includes bibliographical references and index.
  ISBN 978-1-4443-3616-0 (pbk.)
1. Structural engineering.   I. Garrison, Philip, CEng. Basic structures for engineers and architects.   II. Title.
  TA637.G37 2011
  624.1–dc22
                                                                    2010049558

A catalogue record for this book is available from the British Library.

Set in 10/12pt Minion by SPi Publisher Services, Pondicherry, India

1   2011

To my father and late mother, Fred & Jean Garrison

Contents

Introduction

When I was 16 I had a Saturday job as a shelf-stacker at a local supermarket. One day, during a tea break, a co-worker asked me what I did the rest of the week. I explained that I had just done O levels and was going on to do A levels. I told him how many and in which subjects. He then asked me about my career aspirations (not his exact words). I explained that I wanted to become an engineer. His aghast response was: 'What! With all those qualifications?'

Engineers suffer from a lack of public perception of what their profession entails – many people think we spend our days in the suburbs, mending washing machines and televisions. Architects are more fortunate in this respect – the public have a better grasp of their profession: 'They design buildings, don't they?'

Public perceptions aside, careers in both civil engineering and architecture can be extremely rewarding. There are few other careers where individuals can be truly creative, often on a massive scale. The civil engineering profession offers a variety of working environments and a large number of specialisms within civil engineering. Civil engineers have opportunities to work all over the world, on projects large and small, and could come into contact with a wide variety of people, from the lowest worker on a construction site to government officials and heads of state.

In the 21st century there is a huge demand for civil engineers and many young people (and some not so young!) are realising that this is a profession well worth entering.

Traditionally, students embarking on university courses in civil engineering would have A levels in subjects such as mathematics, physics and chemistry. However, for a variety of reasons, many of today's potential students have A levels (or similar) in non-numerate and non-scientific subjects. Moreover, a sizeable number of 'mature' people are entering the profession following a first career in something completely different. As a university admissions tutor, I speak to such people every day. It is possible, depending on the specialism eventually chosen, to enjoy a successful career in civil engineering without an in-depth mathematical knowledge. However, it is extremely difficult to obtain a degree or HND in civil engineering without some mathematical proficiency.

Turning to architects – these are creative people! Every building they design has a structure, without which the building would not stand up. Architects, like civil engineers, have to understand the mechanisms which lead to successful structures.

This book is about Structures. Structures is a subject studied as part of all civil engineering degree, HND and OND courses, as well as architecture degree courses, and also on some degree courses in related subjects (e.g. quantity surveying, building surveying, construction management and architecture).

The purpose of this book

I have taught structures to undergraduate civil engineers and architects for the past 18 years. During that time I have noticed that many students find the basic concept of structures difficult to grasp and apply.

This book aims to do the following:

- to explain structural concepts clearly, using analogies and examples to illustrate the points
- to express the mathematical aspects of the subject in a straightforward manner that can be understood by mathematically weak students and placed in context with the concepts involved
- to maintain reader interest by incorporating into the text real-life examples and case histories to underline the relevance of the material that the student is learning.

This book presumes no previous knowledge of structures on the part of the reader. It does, however, presume that the reader has a good general education and a mathematical ability up to at least GCSE standard.

The intended readership

This book is aimed at:

- National Certificate (ONC), National Diploma (OND), Higher National Certificate (HNC), Higher National Diploma (HND) or first-year degree (BSc, BEng or MEng) students on a civil engineering (or similar) course, who will study a module called structures, structural mechanics, mechanics or structural analysis
- students on a BA degree course in architecture.

The following will also find this book useful:

- students on courses in subjects related to civil engineering and architecture – e.g. quantity surveying, building surveying, construction management or architectural technology – who have to do a structures module as part of their studies
- those studying technology at GCE A level, GNVQ or AVCE
- people working in the construction industry in any capacity.

The following will find the book a useful revision tool:

- a second (or subsequent) year student on a civil engineering or architecture degree
- a professional in the civil engineering or building industry, and practising architects.

A word about computers

Computer packages are available for every specialism, and structural engineering is no exception. Certainly, some of the problems in this book could be solved more quickly using computer software. However, I do not mention specific computer packages in this book and where I mention computers at all, it is in general terms. There are two reasons for this.

1) The purpose of this book is to acquaint the reader with the basic principles of structures. Whereas a computer is a useful tool for solving specific problems, it is no substitute for a thorough grounding in the basics of the subject.

2) Computer software is being improved and updated all the time. The most popular and up-to-date computer package for structural engineering as I write these words may be dated (at best) or obsolete (at worst) by the time you read this. If you are interested in the latest software, look at specialist computer magazines or articles and advertisements in the civil and structural engineering and architecture press, or if you are a student, consult your lecturers.

I have set my students assignments where they have to solve a structural problem by hand then check their results by analysing the same problem using appropriate computer software. If the answers obtained by the two approaches differ, it is always instructive to find out whether the error is in the student's hand calculations (most frequently the case) or in the computer analysis (occurs less frequently, but does happen sometimes when the student has input incorrect or incomplete data – the old 'rubbish in, rubbish out'!).

The website

You will find worked solutions to some of the problems in this book at a website maintained by the publishers: www.blackwellpublishing.com/garrison. In addition, all readers can contact me via the website – your suggestions, comments and criticisms are welcome.

An overview of this book

If you are a student studying a module called structures, structural mechanics or similar, the chapter headings in this book will tie in – more or less – with the lecture topics presented by your lecturer or tutor. I suggest that you read each chapter of this book soon after the relevant lecture or class to reinforce your knowledge and skills in the topic concerned. I advise all readers to have a pen and paper beside them to jot down notes as they go through the book – particularly the numerical examples. In my experience, this greatly aids understanding.

- Chapters 1–5 introduce the fundamental concepts, terms and language of structures.
- Chapters 6–10 build on the basic concepts and show how they can be used, mathematically, to solve simple structural problems.

- Chapter 11 deals with the very important concept of stability and discusses how to ensure structures are stable – and recognise when they're not!
- Chapters 12–15 deal with the analysis of pin-jointed frames, a topic that some students find difficult.
- Chapter 16 covers shear force and bending moment diagrams – an extremely important topic.
- Chapters 17–20 deal with stress in its various guises.
- Structural materials are dealt with more fully in other texts, but Chapter 21 provides an introduction to this topic.
- Chapter 22 has more on materials, and a word on design standards.
- Chapters 23 and 24 deal, respectively, with the conceptual design of structures and the calculation of loads.
- Chapter 25 is a descriptive introduction to structural design, which should be read before embarking on a structural design module.
- Chapter 26 discusses more unusual types of structures.
- Chapter 27 deals with deflection, and outlines a method whereby deflections can be calculated.

How to use this book

It is not necessary for all readers to read this book from cover to cover. However, the book has been designed to follow the subject matter in the order usually adopted by teachers and lecturers teaching structures to students on degree and HND courses in civil engineering. If you are a student on such a course, I suggest that you read the book in stages in parallel with your lectures.

- All readers should read Chapters 1–5 as these lay down the fundamentals of the subject.
- Civil engineering students should read all chapters in the book, with the possible exception of Chapters 14 and 15 if these topics are not taught on your course.
- Students of architecture should concentrate on Chapters 1–9, 21–24 and 26, but read certain other chapters as directed by your tutor.

Let's keep it simple

James Dyson, the inventor of the dual cyclone vacuum cleaner that bears his name, discusses one of its design features – the transparent plastic cylinder within which the rubbish collects – in his autobiography:

A journalist who came to interview me once asked, 'The area where the dirt collects is transparent, thus parading all our detritus on the outside, and turning the classic design inside out. Is this some post-modernist nod to the architectural style pioneered by Richard Rodgers at the Pompidou Centre, where the air-conditioning and escalators, the very guts, are made into a self-referential design feature?

'No,' I replied. 'It's so you can see when it's full.'
(From *Against the Odds* by James Dyson and Giles Coren (Texere 2001))

It is my aim to keep this book as simple, straightforward and jargon-free as possible.

Worked solutions to the tutorial questions can be found at:
www.blackwellpublishing.com/garrison

Acknowledgements

Writing a book is largely an individual, and rather solitary, effort. Every photograph in this book was taken by me, every diagram was drawn by me, and every word of text was written by me on an ageing laptop. However, there have been many who have helped me on the way, including past and present colleagues and family members. My thanks go particularly to the following:

- Katie Bartozzi, Amanda Brown, Andrew Brown, Simon Garrison, Pete Gordon, Paul Hirst and the late Phil Yates.
- Julia Burden and Paul Sayer of Blackwell Publishing.
- My wife Jenny – my greatest fan and my fondest critic.
- Nick Crinson, whose encyclopaedic knowledge of the London Underground inspired the analogy given in Chapter 7, and Maxine Foster, who drew my attention to a discussion of bridge types in a work of popular fiction!
- Colleagues Matt Peat and Dave Roberts who diligently read the first edition of this book and pointed out the several errors that I, the reviewers and proofreaders had all overlooked.
- Brian Walker, who patiently and with great humour taught mathematics to me and my fellow adolescents at Aberdeen Grammar School back in the early 1970s. (If you ever read this Brian do get in touch. Chapter 27 is down to you.)
- Last but not least: Jim Adams, without his inspiration there would have been no book.

And, to parrot that hackneyed catch-all used by all authors at this point, my thanks go to all the others I haven't mentioned, without whom, etc. etc. – they know who they are!

1 What is structural engineering?

Introduction

In this chapter you, the reader, are introduced to structures. We will discuss what a structure actually is. The professional concerned with structures is the structural engineer. We will look at the role of the structural engineer in the context of other construction professionals. We will also examine the structural requirements of a building and will review the various individual parts of a structure and the way they interrelate. Finally you will receive some direction on how to use this book depending on the course you are studying or the nature of your interest in structures.

Structures in the context of everyday life

There is a new confidence evident in major British cities. Redundant Victorian industrial structures are being converted to luxury apartments. Tired old 1960s shopping centres are being razed to the ground, and attractive and contemporary replacements are appearing. Public housing estates built over 40 years ago are being demolished and replaced with more suitable housing. Social shifts are occurring: young professional people are starting to live in city centres and new services such as cafés, bars and restaurants are springing up to serve them. All these new uses require new buildings or converted old buildings. Every building has to have a structure. In some of these new buildings the structure will be 'extrovert' – in other words the structural frame of the building will be clearly visible to passers-by. In many others, the structure will be concealed. But, whether seen or not, the structure is an essential part of any building. Without it, there would be no building.

What is a structure?

The structure of a building (or other object) is the part which is responsible for maintaining the shape of the building under the influence of the forces, loads and other environmental factors to which it is subjected. It is important that the structure as a whole (or any part of it) does not fall down, break or deform to an unacceptable degree when subjected to such forces or loads.

The study of structures involves the analysis of the forces and stresses occurring within a structure and the design of suitable components to cater for such forces and stresses.

Basic Structures, Second Edition. Philip Garrison.
© 2011 John Wiley & Sons, Ltd. Published 2011 by John Wiley & Sons, Ltd.

Fig. 1.1   Lower Manhattan skyline, New York City.
This is one of the largest concentrations of high-rise buildings in the world: space limitations on the island of Manhattan meant that building construction had to proceed upwards rather than outwards, and the presence of solid rock made foundations for these soaring structures feasible.

As an analogy, consider the human body, which comprises a skeleton of 206 bones. If any of the bones in your body were to break, or if any of the joints between those bones were to disconnect or seize up, your injured body would 'fail' structurally (and cause you a great deal of pain).

Examples of structural components (or 'members', as structural engineers call them) include:

- steel beams, columns, roof trusses and space frames
- reinforced concrete beams, columns, slabs, retaining walls and foundations
- timber joists, columns, glulam beams and roof trusses
- masonry walls and columns.

For an example of a densely packed collection of structures, see Fig. 1.1.

What is an engineer?

As mentioned in the introduction, the general public is poorly informed about what an engineer is and what he or she does. 'Engineer' is not the correct word for the person who comes round to repair your ailing tumble drier or office photocopier – nor does it have much to do with engines! In fact, the word 'engineer' comes from the French word *ingénieur*, which refers to someone who uses their ingenuity to solve problems. An engineer, therefore, is a problem-solver.

When we buy a product – for example, a bottle-opener, a bicycle or a loaf of bread – we are really buying a solution to a problem. For instance, you would buy a car not because you wish to have a tonne of metal parked outside your house but rather because of the service it can offer you: a car solves a transportation problem. You could probably think of numerous other examples:

- A can of baked beans solves a hunger problem.
- Scaffolding solves an access problem.
- Furniture polish solves a cleaning problem.
- A house or flat solves an accommodation problem.
- A university course solves an education problem.

A structural engineer solves the problem of ensuring that a building – or other structure – is adequate (in terms of strength, stability, cost, etc.) for its intended use. We shall expand on this later in the chapter. A structural engineer does not usually work alone: he is part of a team of professionals, as we shall see.

The structural engineer in the context of related professions

If I were to ask you to name some of the professionals involved in the design of buildings, the list you would come up with would probably include the following:

- architect
- structural engineer
- quantity surveyor.

Of course, this is not an exhaustive list. There are many other professionals involved in building design (for example, building surveyors and project managers) and many more trades and professions involved in the actual construction of buildings, but for simplicity we will confine our discussion to the three named above.

The architect is responsible for the design of a building with particular regard to its appearance and environmental qualities such as light levels and noise insulation. His starting point is the client's brief. (The client is usually the person or organisation that is paying for the work to be done.)

The structural engineer is responsible for ensuring that the building can safely withstand all the forces to which it is likely to be subjected, and that it will not deflect or crack unduly in use.

The quantity surveyor is responsible for measuring and pricing the work to be undertaken – and for keeping track of costs as the work proceeds.

So, in short:

1) The architect makes sure the building looks good.
2) The (structural) engineer ensures that it will stand up.
3) The quantity surveyor ensures that its construction is economical.

Of course, these are very simplistic definitions, but they'll do for our purposes.

Now I'm not an architect and I'm not a quantity surveyor. (My father is a quantity surveyor, but he's not writing this book.) However, I am a structural engineer and this book is about structural engineering, so in the remainder of this chapter we're going to explore the role of the structural engineer in a little more detail.

Structural understanding

The basic function of a structure is to transmit loads from the point at which the load is applied to the point of support and thus to the foundations in the ground. (We'll be looking at the meaning of the word 'load' more fully in Chapter 5, but for the time being consider a load as being any force acting externally on a structure.)

Any structure must satisfy the following criteria:

1) Aesthetics – it must look nice.
2) Economy – it mustn't cost more than the client can afford; and less if possible.
3) Ease of maintenance.
4) Durability – this means that the materials used must be resistant to corrosion, spalling (pieces falling off), chemical attack, rot and insect attack.
5) Fire resistance – while few materials can completely resist the effects of fire, it is important for a building to resist fire long enough for its occupants to be safely evacuated.

In order to ensure that a structure behaves in this way, we need to develop an understanding and awareness of how the structure works.

Safety and serviceability

There are two main requirements of any structure: it must be **safe** and it must be **serviceable**. 'Safe' means that the structure should not collapse – either in whole or in part. 'Serviceable' means that the structure should not deform unduly under the effects of deflection, cracking or vibration.

Let's discuss these two points in more detail.

Safety

A structure must carry the expected loads without collapsing – either as a whole or even just a part of it. Safety in this respect depends on two factors:

1) The **loading** that the structure is designed to carry has been correctly assessed.
2) The strength of the **materials** that are used in the structure has not deteriorated.

From this it is evident that we need to know how to determine the load on any part of a structure. We will learn how to do this later in the book. Furthermore, we also know that materials deteriorate over time if they are not properly maintained: steel corrodes, concrete may spall or suffer carbonation, and timber will rot. The structural engineer must consider this when designing any particular building.

Serviceability

A structure must be designed in such a way that it doesn't deflect or crack unduly in use. It is difficult or impossible to completely eliminate these things – the important thing is that the deflection and cracking are kept within certain limits. It must also be ensured that vibration does not have an adverse effect on the structure – this is particularly important in parts of buildings containing plant or machinery.

If, when you walk across the floor of a building, you feel the floor deflect or 'give' underneath your feet, it may lead you to be concerned about the integrity of the structure. Excessive deflection does not necessarily mean that the floor is about to collapse, but because it may make people feel unsafe, deflection must be 'controlled'; in other words, it must be kept within certain limits. To take another example, if a lintel above a doorway deflects too much, it may cause warping of the door frame below it and, consequently, the door itself may not open or close properly.

Cracking is ugly and may or may not be indicative of a structural problem. But it may, in itself, lead to problems. For example, if cracking occurs on the outside face of a reinforced concrete wall then rain may penetrate and cause corrosion of the steel reinforcement within the concrete, which in turn leads to spalling of the concrete.

The composition of a building structure

A building structure contains various elements, the adequacy of each of which is the responsibility of the structural engineer. In this section we briefly consider the form and function of each. These elements will be considered in more detail in Chapter 3.

A roof protects people and equipment in a building from weather. An example of a roof structure is shown in Fig. 1.2.

Fig. 1.2   Roof structure of Quartier 206 shopping mall, Berlin. Quite a 'muscular' roof structure!

> If you plan on buying a house in the United Kingdom, be wary of buying one which has a flat roof. Some roofing systems used for waterproofing flat roofs deteriorate over time, leading to leaking and potentially expensive repairs. The same warning applies to flat-roofed additions to houses, such as porches or extensions.

Walls can have one or more of several functions. The most obvious one is **load-bearing** – in other words, supporting any walls, floors or roofs above it. But not all walls are load-bearing. Other functions of a wall include:

- partitioning, or dividing, rooms within a building – and thus defining their shape and extent
- weatherproofing
- thermal insulation – keeping heat in (or out)
- noise insulation – keeping noise out (or in)
- fire resistance
- security and privacy
- resisting lateral (horizontal) loads such as those due to retained earth, wind or water.

Consider the wall closest to you as you read these words. Is it likely to be load-bearing? What other functions does this wall perform?

A *floor* provides support for the occupants, furniture and equipment in a building. Floors on an upper level of a building are always *suspended*, which means that they span between supporting walls or beams. Ground floor slabs may sit directly on the ground beneath.

Staircases provide for vertical movement between different levels in a building. Figure 1.3 shows a concrete staircase in a multi-storey building. Unusually, the staircase is fully visible from outside the building. How is this staircase supported structurally?

Foundations represent the interface between the building's structure and the ground beneath it. A foundation transmits all the loads from a building into the ground in such a way that settlement (particularly uneven settlement) of the building is limited, and failure of the underlying soil is avoided.

> On a small sandy island in the Caribbean, a low-rise hotel was being constructed as part of a larger leisure resort. The contractor for the hotel (a somewhat maverick individual) thought he could save money by not constructing foundations. He might have got away with it were it not for an alert supervising engineer, who spotted that the blockwork walls did not appear to be founded on anything more rigid than sand.
>
> A furious argument ensued between the design team and the contractor, who not only readily admitted that no foundations had been built but also asserted that, in his opinion, none were required. In a developed country the contractor would have been dismissed instantly and probably prosecuted, but things were a little more free and easy in this corner of the Caribbean.
>
> But nature exacted its own retribution. That night, a tropical storm blew up, the sea washed over the island … and the partly built structure was entirely washed away.

Fig. 1.3 **A very visible staircase. How is it supported? You'll learn more about cantilevers later in the book.**

Fig. 1.4 **A conventional building enclosed in a glazed outer structure. The two structures appear to be completely independent of each other.**

In a building it is frequently necessary to support floors or walls without any interruption or division of the space below. In this case, a horizontal element called a **beam** will be used. A beam transmits the loads it supports to columns or walls at the beam's ends.

A **column** is a vertical load-bearing element which usually supports beams and/or other columns above. Laymen often call them pillars or poles or posts. Individual elements of a structure, such as beams or columns, are often referred to as **members**.

See Fig. 1.4 for an unusual pairing of two separate structures.

A few words for students on architecture courses

If you are studying architecture, you may be wondering why you need to study structures at all. It is not the purpose of this book to make you a fully qualified structural engineer, but, as an architect, it is important that you understand the principles of structural behaviour. Moreover, with some basic training there is no reason why architects cannot design simple structural members (e.g. timber joists supporting floors) themselves. On larger projects architects work in inter-disciplinary teams which usually include structural engineers. It is therefore important to understand the role of the structural engineer and the language and terms that the structural engineer uses.

How does the study of structures impinge on the training of an architect?

If you are on a degree course in architecture you will have formal lectures in structures throughout your course. You will also be assigned projects involving the architectural design of buildings to satisfy given requirements. It is essential to realise that all parts of the building need to be supported. Always ask yourself the question: 'How will my building stand up?' Remember – if you have difficulty in getting a model of your building to stand up, it is unlikely that the real thing will stand up either!

2 Learn the language: a simple explanation of terms used by structural engineers

Introduction

Structural engineers use the following words (amongst others, of course) in technical discussions:

- force
- reaction
- stress
- moment

None of these words is new to you; they are all common English words that are used in everyday speech. However, in structural engineering each of these words has a particular meaning. In this chapter we shall have a brief look at the specific meanings of the above words before exploring them in more detail in later chapters.

Force

A force is an ***influence*** on an object (for example, part of a building) that may cause movement. For example, the weight of people and furniture within a building causes a vertically downwards force on the floor, and wind blowing against a building causes a horizontal (or near horizontal) force on the external wall of the building.

Force is discussed more fully in Chapter 4, together with related terms such as mass and weight. Forces are also sometimes referred to as ***loads*** – the different types of load are reviewed in Chapter 5.

Reaction

If you stand on a floor (or a roof! – see Fig. 2.1), the weight of your body will produce a downward force into the floor. The floor reacts to this by pushing upwards with a force of the same magnitude as the downward force due to your body weight. This upward force is called a ***reaction***, as its very presence is a response to the downward force of your body. Similarly, a wall or a column supporting a beam will produce an upward reaction as a response to the downward forces the beam transmits to the wall (or column) and a foundation will produce an upward reaction to the downward force in the column or wall that the foundation is supporting.

Basic Structures, Second Edition. Philip Garrison.
© 2011 John Wiley & Sons, Ltd. Published 2011 by John Wiley & Sons, Ltd.

Fig. 2.1   Oslo Opera House.
Floors don't always have to be flat! At Oslo's new opera house the public is encouraged to walk all over its sloping roof.

The same is true of horizontal forces and reactions. If you push horizontally against a wall, your body is applying a horizontal force to the wall – which the wall will oppose with a horizontal reaction.

The concept of a reaction is discussed in more detail in Chapter 6 and you will learn how to calculate reactions in Chapter 9.

Stress

Stress is internal pressure. A heavy vehicle parked on a road is applying pressure to the road surface – the heavier the vehicle and the smaller the contact area between the vehicle's tyres and the road, the greater the pressure. As a consequence of this pressure on the road surface, the parts of the road below the surface will experience a pressure which, because it is within an object (in this case, the road) is termed a *stress*. Because the effect of the vehicle's weight is likely to be spread, or dispersed, as it is transmitted downwards within the road structure, the stress (internal pressure at a point) will decrease the further down you go within the road's construction.

So, stress is ***internal pressure*** at a given point within, for example, a beam, slab or column. It is likely that the intensity of the stress will vary from point to point within the object.

Stress is a very important concept in structural engineering. In Chapters 17–20, you will learn more about how to calculate stresses.

Moment

A moment is a turning effect. When you use a spanner to tighten a nut, mechanically wind up a clock or turn the steering wheel on your car, you are applying a moment. The concept and calculation of moments is discussed in Chapter 8.

The importance of 'speaking the language' correctly

A major American bank planned changes to its London headquarters building that entailed the removal of substantial internal walls. Although a well-known firm of structural engineers was used for the design, the work itself was entrusted to a firm of shopfitters who clearly had no experience whatsoever in this type of work.

The client issued the structural engineer's drawings to the shopfitting contractor. In a site meeting, the contractor asked the structural engineer if it would be all right to use steel 'H' sections at the points where 'UC' columns were indicated on the drawings. The structural engineer was a little puzzled by this and pointed out that 'UC' stands for universal column, which are indeed steel 'H' sections. The contractor admitted, a little sheepishly, that he had thought that 'UC' stood for 'U-shaped channel section'!

The structural engineer was so shaken by this conversation and its potential consequences that he strongly advised the client to sack the shopfitters and engage contractors who knew what they was doing.

3 How do structures (and parts of structures) behave?

Introduction

In this chapter we will discuss how parts of a structure behave when they are subjected to forces. We will consider the meanings of the terms **compression**, **tension**, **bending** and **shear**, with examples of each. Later in the chapter we will look at the various elements that make up a structure, and at different types of structure.

Compression

Figure 3.1a shows an elevation – that is, a side-on view – of a concrete column in a building. The column is supporting beams, floor slabs and other columns above, and the load, or force, from all of these is acting downwards at the top of the column. This load is represented by the downward arrow at the top of the column. Intuitively, we know that the column is being squashed by this applied load – it is experiencing compression.

As we have seen briefly in Chapter 2 and will discuss more fully in Chapter 6, a downward force must be opposed by an equal upward force if the building is stationary – as it should be. This reaction is represented by the upward arrow at the bottom of the column in Fig. 3.1a. Now, not only must the rules of equilibrium (total force up = total force down) apply for the column as a whole; these rules must apply *at any and every point* within a stationary structure.

Let's consider what happens at the top of the column – specifically, point C in Fig. 3.1b. The downward force shown in Fig. 3.1a at point C must be opposed by an upward force – also at point C. Thus there will be an upward force within the column at this point, as represented by the upward broken arrow in Fig. 3.1b. Now let's consider what happens at the very bottom of the column – point D in Fig. 3.1b. The upward force shown in Fig. 3.1a at point D must be opposed by a downward force at the same point. This is represented by the downward broken arrow in Fig. 3.1b.

Look at the direction of the broken arrows in Fig. 3.1b. These arrows represent the internal forces in the column. You will notice that they are pointing away from each other. This is always the case when a structural element is in compression: the arrows used to denote compression *point away* from each other.

Basic Structures, Second Edition. Philip Garrison.
© 2011 John Wiley & Sons, Ltd. Published 2011 by John Wiley & Sons, Ltd.

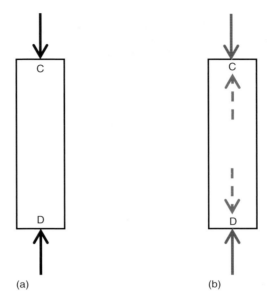

Fig. 3.1   **A column in compression.**

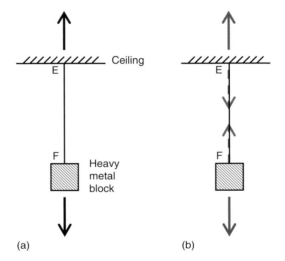

Fig. 3.2   **A piece of string in tension.**

Tension

Figure 3.2 shows a heavy metal block suspended from the ceiling of a room by a piece of string. The metal block, under the effects of gravity, is pulling the string downwards, as represented by the downward arrow. The string is thus being stretched and is therefore in tension.

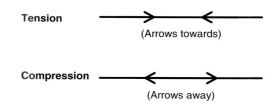

Fig. 3.3   Arrow notations for the internal forces of tension and compression.

For equilibrium, this downward force must be opposed by an equal upward force at the point where the string is fixed to the ceiling. This opposing force is represented by an upward arrow in Fig. 3.2a. Note that if the ceiling wasn't strong enough to carry the weight of the metal block, or the string was improperly tied to it, the weight would come crashing to the ground and there would be no upward force at this point. As with the column considered above, the rules of equilibrium (total force up = total force down) must apply at any and every point within this system if it is stationary.

Let's consider what happens at the top of the string. The upward force shown in Fig. 3.2a at point E must be opposed by a downward force – also at this point. Thus there will be a downward force within the string at this point, as represented by the downward broken arrow in Fig. 3.2b. Now let's consider what happens at the very bottom of the string – at the point where the metal block is attached (point F). The downward force shown in Fig. 3.2a at point F must be opposed by an upward force at this point. This upward force within the string at this point is represented by the upward broken arrow in Fig. 3.2b.

Look at the direction of the broken arrows in Fig. 3.2b. These arrows represent the internal forces in the string. You will notice that they are pointing towards each other. This is always the case when a structural element is in tension: the arrows used to denote tension *point towards* each other. (An easy way to remember this principle is the letter T, which stands for both Towards and Tension.)

The standard arrow notations for the internal forces of members in tension and compression are shown in Fig. 3.3. You should familiarise yourself with them as we shall meet them again in later chapters.

Note: Tension and compression are both examples of *axial* forces – they act along the *axis* (or centre line) of the structural member concerned.

Bending

Consider a simply supported beam (that is, a beam that simply rests on supports at its two ends) subjected to a central point load. The beam will tend to bend, as shown in Fig. 3.4. The extent to which the beam bends will depend on four things:

1) The *material* from which the beam is made. You would expect a beam made of rubber to bend more than a concrete beam of the same dimensions under a given load.
2) The *cross-sectional characteristics* of the beam. A large diameter wooden tree trunk is more difficult to bend than a thin twig spanning the same distance.

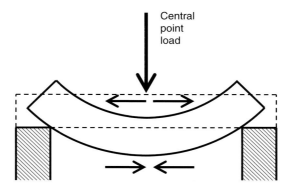

Fig. 3.4   **Bending in a beam.**

3)   The *span* of the beam. Anyone who has ever tried to put up bookshelves at home will know that the shelves will sag to an unacceptable degree if not supported at regular intervals. The same applies to the hanger rail inside a wardrobe. The rail will sag noticeably under the weight of all those clothes if it is not supported centrally as well as at its ends, thereby reducing the span.

4)   The *load* to which the beam is subjected. The greater the load, the greater the bending. Your bookshelves will sag to a greater extent under the weight of heavy encyclopedias than they would under the weight of a few light paperback books.

If you carry on increasing the loading, the beam will eventually break. Clearly, the stronger the material, the more difficult it is to break. A wooden ruler is quite easy to break by bending; a steel ruler of similar dimensions might bend quite readily but it's unlikely that you would manage to break it with your bare hands!

This is evidently one way in which a beam can fail – through excessive bending. Beams must be designed so that they do not fail in this way.

Incidentally, one way of determining the amount of bending that has taken place is to calculate the ***deflection***, which is the vertical distance through which a given point on the beam has moved from its original position. We'll discuss deflection later in the book.

Shear

Consider two steel plates that overlap each other slightly, with a bolt connecting the two plates through the overlapping part, as shown in Fig. 3.5a. Imagine now that a force is applied to the top plate, trying to pull it to the left. An equal force is applied to the bottom plate, trying to pull it to the right. Let's now suppose that the leftward force is slowly increased, as is the rightward force. (Remember that the two forces must be equal if the whole system is to remain stationary.) If the bolt is not as strong as the plates, eventually we will reach a point when the bolt will break. After the bolt has broken, the top part of it will move off to the left with the top plate and the bottom part will move off to the right with the bottom plate.

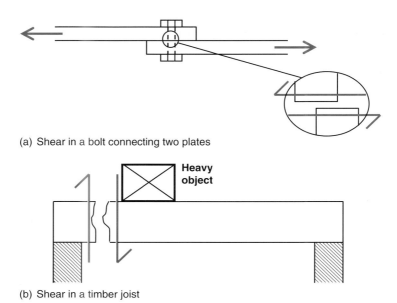

(a) Shear in a bolt connecting two plates

(b) Shear in a timber joist

Fig. 3.5 The concept of shear.

Let's examine in detail what happens to the failure surfaces (that is, the bottom face of the top part of the bolt and the top face of the bottom part of the bolt) immediately after failure. As you can see from the 'exploded' part of Fig. 3.5a, the two failure surfaces are sliding past each other. This is characteristic of a shear failure.

We'll now turn our attention to a timber joist supporting the first floor of a building, as shown in Fig. 3.5b. Let's imagine that timber joists are supported on masonry walls and that the joists themselves support floorboards, as would be the case in a typical domestic dwelling – such as, perhaps, the house you live in. Suppose that the joists are inappropriately undersized – in other words, they are not strong enough for the loads they are likely to have to support.

Now let's examine what would happen if a heavy object – for example, some large piece of machinery – was placed on the floor near its supports, as shown in Fig. 3.5b. If the heavy object is near the supporting walls, the joists may not bend unduly. However, if the object is heavy enough and the joists are weak enough, the joist may simply break. This type of failure is analogous to the bolt failure discussed above. With reference to Fig. 3.5b, the right-hand part of the beam will move downwards (as it crashes to the ground), while the left-hand part of the beam will stay put – in other words, it moves upwards relative to the downward-moving right-hand part of a beam. So, once again, we get a failure where the two failure surfaces are sliding past each other: a shear failure. So a shear failure can be thought of as a cutting or slicing action.

So, this is a second way in which a beam can fail – through shear. Beams must be designed so that they do not fail in this way. (Incidentally, the half-headed arrow notation shown in Fig. 3.5 is the standard symbol used to denote shear.)

The consequences of bending and shear failures – and how to design against them – will be discussed more fully in Chapter 16.

Structural elements and their behaviour

The various types of structural element that might be found in a building – or any other – structure were introduced in Chapter 1. Now we've learned about the concepts of compression, tension, bending and shear, we'll discuss how these different parts of a structure behave under load.

Beams

Beams may be *simply supported*, *continuous* or *cantilevered*, as illustrated in Fig. 3.6. They are subjected to bending and shear under load, and the deformations under loading are shown by broken lines.

A simply supported beam rests on supports, usually located at each end of the beam. A continuous beam has two or more spans in one unbroken unit; it may simply rest on its supports, but more usually it is gripped (or fixed) by columns above and below it. A cantilever beam is supported at one end only, so to avoid collapse, the beam must be continuous over, or rigidly fixed at, this support.

Beams may be of timber, steel or reinforced or prestressed concrete.

Slabs

As with beams, slabs span horizontally between supports and may be simply supported, continuous or cantilevered. But unlike beams, which are usually narrow compared with their depth, slabs are usually wide and relatively shallow and are designed to form flooring – see Fig. 3.7.

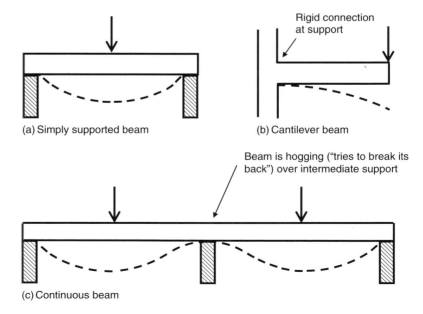

Fig. 3.6 **Beam types.**

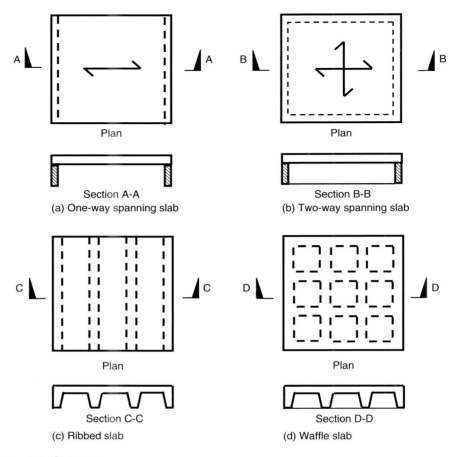

Fig. 3.7   **Slab types.**

Slabs may be one-way spanning, which means that they are supported by walls on opposite sides of the slab, or two-way spanning, which means that they are supported by walls on all four sides. This description assumes that a slab is rectangular in plan, as is normally the case. Slabs are usually of reinforced concrete and in buildings they are typically 150–300 millimetres in depth. Larger than normal spans can be achieved by using ribbed or waffle slabs, as shown in Fig. 3.7c and d. Like beams, slabs experience bending.

Columns

Columns (or 'pillars' or 'posts') are vertical and support axial loads, thus they experience compression. If a column is slender or supports a non-symmetrical arrangement of beams, it will also experience bending, as shown by the broken line in Fig. 3.8a. Concrete or masonry columns may be of square, rectangular, circular or cruciform cross-section, as illustrated in Fig. 3.8b. Steel columns may be H or hollow section, as illustrated later in this chapter.

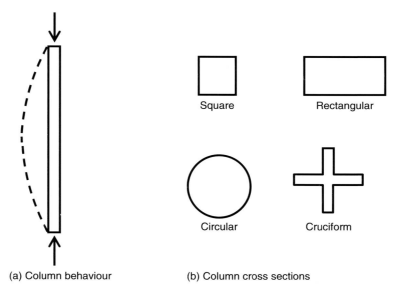

(a) Column behaviour   (b) Column cross sections

Fig. 3.8   Column types.

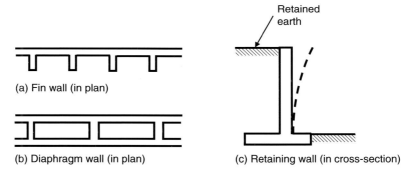

(a) Fin wall (in plan)

(b) Diaphragm wall (in plan)   (c) Retaining wall (in cross-section)

Fig. 3.9   Wall types.

Walls

Like columns, walls are vertical and are primarily subjected to compression, but they may also experience bending. Walls are usually of masonry or reinforced concrete. As well as conventional flat-faced walls you might encounter fin or diaphragm walls, as shown in Fig. 3.9. *Retaining walls* hold back earth or water and thus are designed to withstand bending caused by horizontal forces, as indicated by the broken line in Fig. 3.9c.

Foundations

As mentioned in Chapter 1, everything designed by an architect or civil or structural engineer must stand on the ground – or at least have some contact with the ground.

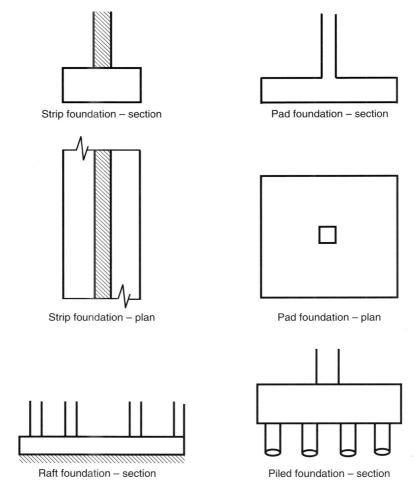

Strip foundation – section

Pad foundation – section

Strip foundation – plan

Pad foundation – plan

Raft foundation – section

Piled foundation – section

Fig. 3.10 Foundation types.

So they need foundations, whose function is to transfer loads from the building safely into the ground. There are various types of foundation. A *strip* foundation provides a continuous support to load-bearing external walls. A *pad* foundation provides a load-spreading support to a column. A *raft* foundation takes up the whole plan area under a building and is used in situations where the alternative would be a large number of strip and/or pad foundations in a relatively small space. Where the ground has low strength and/or the building is very heavy, *piled* foundations are used. These are columns in the ground which transmit the building's loads safely to a stronger stratum. A group of piles is topped by a large concrete block called a pile cap. All these foundation types are illustrated in Fig. 3.10, and a pile cap under construction is shown in Fig. 3.11.

Foundations of all types are usually of concrete, but occasionally steel or timber may be used for piles.

Fig. 3.11 Pile cap under construction.
The steel reinforcement cages have been placed, but the concrete has not yet been poured; the bright yellow plastic clearly indicates the outline of the pile caps – a large one behind the crane, with much smaller ones to the right.

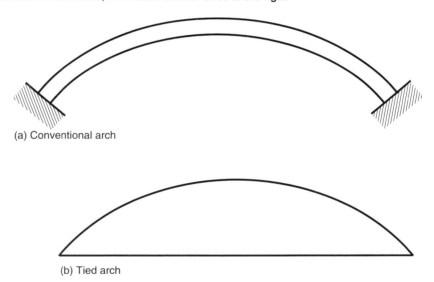

(a) Conventional arch

(b) Tied arch

Fig. 3.12 Arch types.

Arches

The main virtue of an arch, from a structural engineering point of view, is that it is in compression throughout. This means that materials that are weak in tension – for example, masonry – may be used to span considerable distances. Arches transmit large horizontal thrusts into their supports, unless horizontal ties are used at the base of the arch. It is to cope with these horizontal thrusts that flying buttresses are provided in medieval cathedrals – see Fig. 3.12. You can read more about arches in Chapter 26.

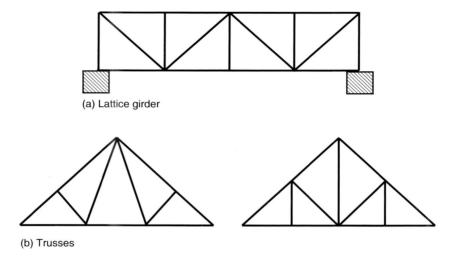

(a) Lattice girder

(b) Trusses

Fig. 3.13   Truss types.

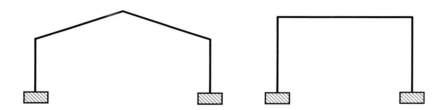

Fig. 3.14   Portal frame types.

Trusses

A truss is a two- or three-dimensional framework and is designed on the basis that each 'member' or component of the framework is in either pure tension or pure compression and does not experience bending. Trusses are often used in pitched roof construction: timber tends to be used for domestic construction and steel caters for the larger roof spans required in industrial or commercial buildings. Lattice girders, which are used instead of solid deep beams for long spans, work on the same principle – see Fig. 3.13.

Portal frames

A portal frame is a rigid framework comprising two columns supporting rafters. The rafters may be horizontal or, more usually, inclined to support a pitched roof. Portal frames are usually of steel but may be of precast concrete. They are usually used in large single-storey structures such as warehouses or out-of-town retail sheds – see Fig. 3.14.

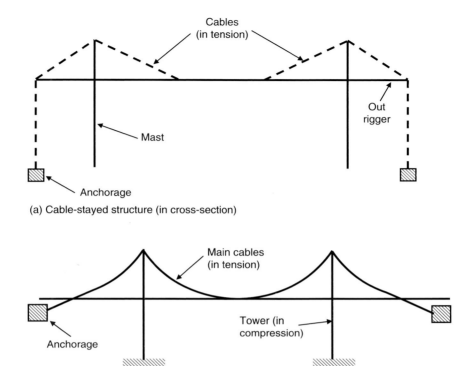

Fig. 3.15　**Cable stayed and suspension structures.**

Cable-stayed and suspension structures

Cable-stayed structures are usually bridges but they are sometimes used in building structures where exceptionally long spans are required. Instead of being supported from below by columns or walls, the span is supported from above at certain points by cables that pass over supporting vertical masts and horizontal outriggers to a point in the ground where they are firmly anchored. A similar principle applies to suspension bridges. The cables are in tension and must be designed to sustain considerable tensile forces – see Fig. 3.15. Examples of suspension bridges are shown in Figures 3.16 and 3.17.

Cross-sectional types

There is a large range of cross-sectional shapes available. Standard sections are illustrated in Fig. 3.18.

- Beams and slabs in timber and concrete are usually rectangular in cross-section.
- Concrete columns are usually of circular, square, rectangular or cruciform cross-section (see above).

- Steel beams are usually of 'I' or hollow section.
- Steel columns are usually of 'H' or hollow section.
- Prestressed concrete beams are sometimes of 'T', 'U' or inverted 'U' section.
- Members of steel trusses are sometimes of channel or angle sections.
- Steel Z purlins (not illustrated) are often used to support steel roofing or cladding.

Fig. 3.16   Brooklyn Bridge, New York City.
Conceived by John Roebling and completed by his son, Washington, in 1883, the Brooklyn Bridge was the first suspension bridge in the world to use steel for its main cables and was the longest suspension bridge in the world at the time of its construction; the foundations were excavated from within underwater caissons that used compressed air, causing crippling illnesses among the workers, including Washington Roebling himsel.

Fig. 3.17   Millennium Footbridge and Tate Modern, London.
Two contrasting structures: a modern low-slung pedestrian suspension bridge leads to a 1950s brick power station converted into a modern art gallery.

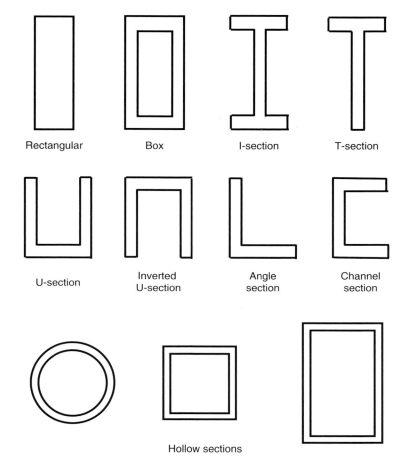

Fig. 3.18   Cross section types.

Appraisal of existing structures

Steam room indiscretion

One Saturday morning I was relaxing in the steam room at my local fitness centre after a punishing workout. My companions there were two men in their 20s and an older man, possibly mid 40s. All were sat in their swimming trunks, happily sweating away.

The two younger men were clearly friends. As I entered the steam room, they were in the middle of a conversation about a forthcoming party, which went something like this:

MAN 1:   Is Craig going?
MAN 2:   Yes, I think so.
MAN 1:   Oh, good – he's a real laugh.
MAN 2:   Yes – he's a nutcase. (Pause) Craig was leaving the rugby club last Sunday lunch-time after a few drinks when he smashed his car up. So he pushed it off the road into a field, covered it with straw, then phoned the police and reported it stolen.

MAN 1:   (After a slight pause) Did he get away with it?
MAN 2:   Er – yes, I think so.

The older man, who'd remained silent up to this point, now spoke up slowly and deliberately: 'The thing that you've got to remember is that when you're in a steam room, police officers are not wearing uniforms.'
There was an uncomfortable silence while the significance of this remark sank in. The police officer eventually smoothed things over by telling a similar anecdote of his own, after which he left the steam room. One of the young men turned to me and his friend and said: 'Well, that could have been a bit unfortunate, couldn't it!'
'Yes,' I replied, 'I'm CID.'

The story above has a serious message. People are not always what they seem: a near-naked police officer looks much the same as anyone else. Similarly, structures are not always what they might seem either, although the problem here is usually one of too much cladding. In some buildings the designers choose to make a feature of the structure; in others, the structure is totally concealed. You may hear architects (and others) describe a given building as "structurally extrovert", meaning that the structure is clearly visible and perhaps visually dominant.

In your future professional career you may be called upon to carry out a structural inspection of an existing building, usually after someone else – perhaps a building surveyor – has identified a fault that they suspect may be structural in nature. It is not always easy to assess how an existing building functions structurally. Certainly, you can pick up clues from the age and style of the building, and original drawings of the structure as built are very useful – though not always reliable – in the unlikely event that they are available at all.

So, if you have to carry out a structural appraisal of an existing building, my advice is to tread carefully.

What you should remember from this chapter

The concepts of compression, tension, bending and shear are fundamental to any study of structural mechanics. The reader should clearly understand the meaning and implications of each. Different elements of a structure deform in different ways under load. The reader should understand and be able to visualise these patterns of structural behaviour, which are fundamental to structural design.

4 Force, mass and weight

Introduction

In this chapter we look at force, mass and weight – their definitions, the relationships between them, their units of measurement and their practical application.

Force

We use the term 'force' in everyday life. For example, somebody may force you to do something. This means that that person, through their words, actions or other behaviour, compels you to take a certain course of action. The word force in a technical context is similar: a *force* is an influence, or action, on a body or object which causes – or attempts to cause – movement. For example:

- The man shown in Fig. 4.1 is pushing against a wall. In doing so, he is applying a horizontal force to that wall – in other words, he is attempting to push that wall away from him.
- Figure 4.2 shows a man standing on a hard surface. The weight of his body is applying a vertical (downwards) force on the floor – in other words, he is attempting to move the floor downwards.

Force is measured in units of newtons (N) or kilonewtons (kN) – but more of that later.

Mass

Mass is the amount of *matter* in a body or object. It is measured in units of grams (g) or kilograms (kg). Mass should not be confused with weight (see below).

Weight

If you have studied physics, or science generally, you will have come across the following equation:

$$\text{Force} = \text{Mass} \times \text{Acceleration}$$

Basic Structures, Second Edition. Philip Garrison.
© 2011 John Wiley & Sons, Ltd. Published 2011 by John Wiley & Sons, Ltd.

Fig. 4.1   Man leaning against a wall

A much more useful form of this equation to engineers is:

Weight = Mass × Acceleration due to Gravity

If an object is dropped from a height, it will accelerate – that is, consistently increase in speed – as it heads towards the ground. This acceleration is called the acceleration due to gravity and its value at the surface of this planet is 9.81 metres/sec². This means that a body drop will, after one second, be falling at 9.81 m/s, and after two seconds will have reached (9.81 + 9.81) = 19.62 m/s. etc. The mass of the object is irrelevant in this context – a bundle of feathers falls at the same rate as a large lump of lead, as they experience the same rate of acceleration.

Looking at the above equation, we see that the relationship between mass and weight is governed by the acceleration due to gravity. It suggests that an object of a given mass will weigh less on a planet where the gravitational pull is less. If you watch television footage of the Apollo moon landings in the late 1960s, you will notice that the astronauts appear to be leaping and bounding around on the moon in a manner that would be regarded as undignified on earth. This is because although the mass of a particular astronaut (that is, the amount of matter in his body) is obviously the same on the moon as it is on earth, the gravitational pull (and hence acceleration due

Fig. 4.2   Man standing on a floor

to gravity) is much less on the moon and therefore the astronaut's weight is much less on the moon than it is on earth.

The relationship between weight and mass

Coming back to earth, if an object of mass 1 kg is subjected to the acceleration due to gravity of 9.81 m/sec^2 (which is approximately 10 m/sec^2), then the above equation tells us that the object's weight is $(1 \times 10) = 10$ N. Note that weight is a force and is measured in the same units as force: newtons (N) or kilonewtons (kN). So:

A *weight* of 10 N is equivalent, at the earth's surface, to a *mass* of 1 kg

As you are probably aware, the kilo- prefix means '1000 times', so:

1000 N = 1 kN

Therefore, if a weight of 10 N is equivalent to a mass of 1 kg, then a weight of 1000 N (or 1 kN) is equivalent to a mass of 100 kg.

One further relationship: a weight of 10 kN is equivalent to a tonne (also known as a metric tonne), i.e. 1000 kg.

What do these units signify in everyday terms?

1) Sugar is sold in your local supermarket in 1 kg bags. If you lift a 1 kg bag of sugar, you will get some idea of what a mass of 1 kg (and hence a force of 10 N) feels like.

2) 1 kN is equivalent to 100 kg, which in turn is approximately 220 pounds or just under 16 stone – 16 stone is the weight of a reasonably large man. If you imagine a large man of your acquaintance, then the mass of their body is imposing a 1 kN force downwards.

3) A small modern car weighs about a tonne, so has a weight of about 10 kN.

To summarise:

- 10 N is the weight of 1 kg (a bag of sugar)
- 1000 N or 1 kN is the weight of a 100 kg (16-stone) person
- 10 kN is the weight of 1000 kg = 1 tonne (a small car)

Density and unit weight

The *density* of a material can be calculated as follows:

$$\text{Density (kg / m}^3) = \frac{\text{Mass (kg)}}{\text{Volume (m}^3)}$$

Unit weight is a similar concept to density. The unit weight is the weight of a material per unit volume and is measured in kN/m³. Unit weights of some common building materials are given in Appendix 1.

Units generally

You should always be conscious of the units you are using in any structural calculation. Incorrect use and understanding of units can lead to wildly inaccurate answers.

The lecturers and tutors who mark and assess your coursework and examinations are well aware of the perils of getting the units wrong. Make sure that you express the units in any written work you do. For example, the force in a column is not 340, but it might be 340 kN. Omitting units is sheer laziness and it may lead the person assessing your work to doubt whether you understand what you're doing and they will mark the work accordingly.

Relationships with other measuring systems

Although the metric system is now generally used in scientific and technical work in the UK, you will need to know how to convert from the Imperial system of measurement – pounds, feet, inches, etc. This is because, in your future professional career:

- you may have to review calculations or drawings made before the 1960s when the metric system came into use
- you may be working in (or for) a country which doesn't use the metric system
- you may be dealing with a profession that feels more comfortable with non-metric units – for example, property professionals in the UK still often measure room areas in square feet rather than square metres.

For example:

- 1 pound = 0.454 kg
- 1 inch = 25.4 millimetres

For a more comprehensive list of conversions between different systems of measurement, see Appendix 2.

What you should remember from this chapter

- Mass is the amount of matter in an object and is measured in grams (g) or kilograms (kg).
- Weight is a *force* and is measured in newtons (N) or kilonewtons (kN).
- Density is the ratio of mass to volume and is measured in kg/m³.
- Unit weight is the weight of a material per unit volume and is measured in kN/m³.
- In any calculations in structures the units used should always be expressed.

Tutorial examples

Answers are given at the end of the chapter.

1) Calculate the weight, in kN, of each of the following two people:
   a) A young woman with a mass of 70 kg.
   b) A middle-aged man with a mass of 95 kg.
   What would be the weights of each of these people on the moon if the gravitational acceleration on the moon is one-sixth of that on earth?
2) Calculate the mass of a brick of length 215 mm, breadth 102.5 mm and height 65 mm, if its density is 1800 kg/m³. What would be the weight of this brick?
3) Calculate the weight of a 9-metre-long reinforced concrete beam of breadth 200 mm and depth 350 mm if the unit weight of reinforced concrete is 24 kN/m³.
4) As we will see in later chapters, the term *live load* is used to describe non-permanent load within a building – that is, those loads due to people and furniture. If a university classroom is 12 metres long and 10 metres wide and is designed to accommodate up to 60 students, calculate the live load in the classroom when full. (Note that you will have to make an assessment of the weight of an individual student, desk and chair.) Compare your answer with the British Standard value of live load (3.0 kN/m²) for classrooms.

5) An international hotel chain plans to upgrade its hotel in a particular glamorous and exotic location by installing a rooftop swimming pool on top of its existing high-rise bedroom block. The swimming pool will be 25 metres long and 10 metres wide and will vary uniformly in depth from 1 metre to 2 metres. Calculate the volume of water in the pool. If the unit weight of water is $10\,\text{kN/m}^3$, calculate the mass of water in the pool, in tonnes. If a small modern car has a mass of 1 tonne, calculate the number of cars that would be equivalent, in weight, to the water in the proposed swimming pool. If you were appointed as structural engineer for the project, what would be your initial advice to the architect and client?

6) You are involved in a housing development project. You measure the site on a plan and find that it is rectangular, of length 300 metres and width 250 metres. 'What's the area in acres?' the developer asks you. What is your reply? (Hint: refer to Appendix 2.)

7) You were delighted to win the full £1 million prize money during your recent appearance on a television quiz programme. However, your elation abates somewhat when the programme's producer informs you that the prize money will be given to you in cash, entirely in pound coins. Calculate the mass, weight and volume of 1 million pound coins.

Bearing in mind the sudden interest shown in you by several tabloid newspapers, explain how you would transport the cash from the television studio to your home 200 miles away.

Noting your concerns, the producer offers to pay your prize money in £2 coins instead. Calculate the appropriate mass, weight and volume for this case. Would you accept or decline this offer?

On your eventual arrival at home with your haul, you decide to store the money in a spare bedroom. Assuming conventional timber joist floor construction, do you think this would pose a structural problem and why?

Tutorial answers

1) a) $0.7\,\text{kN}$; b) $0.95\,\text{kN}$. On moon: a) $0.117\,\text{kN}$; b) $0.158\,\text{kN}$.
2) $2.52\,\text{kg}$; $0.025\,\text{kN}$.
3) $15.1\,\text{kN}$.
4) Your answer will probably be in the range $0.5\text{–}1.0\,\text{kN/m}^2$, depending on your assumptions.
5) $375\,\text{m}^3$; 375 tonnes.
6) 18.5 acres.

5 Loading – dead or alive

Introduction: what is a load?

As discussed in Chapter 2, a **load** is a force on a part of a structure. The term 'load' is frequently used in everyday life. We refer to 'loading' a washing machine when we fill it with clothes to be washed. You 'load up' your car before going on a motoring holiday. The airline industry uses the term 'load factors' when describing how many passengers get on flights, and an insurance company will 'load' your premium (in other words, increase the amount you have to pay for your insurance policy) if you give information which it feels increases the risk of it having to pay out. In short, you are already familiar with the word load, and its use in structural engineering is, I hope, easy to understand.

In structures, we have to deal with three different types of loading: dead (or permanent) loads, live (or imposed) loads and wind (or lateral) loads.

Dead (permanent) load

As its alternative name suggests, a dead load is something that is always present. Examples of dead load include the loads – or forces – due to the weights of the various elements of construction, such as floors, walls, roofs, cladding and permanent partitions. These items – and their weights – are obviously always there, 24/7.

Live (imposed) load

Live loads are not always present. They are produced by the **occupancy** of the building. Examples of live loads include people and furniture. Other examples include snow loads on roofs. By their very nature, live loads, unlike dead loads, are variable. For example, a 300-seat cinema auditorium would be full of people on a Saturday evening if a major new blockbuster movie was being shown, but it might be less than a quarter full on a weekday afternoon. And, of course, it would be empty when the cinema is closed. So the live load in this cinema auditorium is represented by anything between 0 and 300 people. A classroom in a college or university is a similar example. The classroom might be full of students or empty – or anything in between. Also, it might be decided to temporarily remove the desks and chairs from the classroom – to hold an exhibition there, for example.

Basic Structures, Second Edition. Philip Garrison.
© 2011 John Wiley & Sons, Ltd. Published 2011 by John Wiley & Sons, Ltd.

Fig. 5.1   London Eye.
**Built to celebrate the millennium London Eye – designed by Marks Barfield Architects –
is a steel and glass Ferris wheel structure resembling a giant bicycle wheel, and many
structural and logistical problems had to be overcome in its design and construction.**

Because of this variability, live loads are treated differently from dead loads in
structural design.

A live load does not have to be moving, animate or alive in any way. For example,
a dead body in a mortuary is a live load because it is there only temporarily. A car in a
multi-storey car park will be a live load whether or not it is moving: again, the assump-
tion is that the car is there for only a certain period of time and then it will be removed.

In short, dead loads are there all the time, live loads are not.

Wind (lateral) load

Wind loading is an example of lateral loading. Unlike dead and live loads, which are
usually vertical in direction, wind loads act horizontally or at a shallow angle to the
horizontal. Wind loads vary across the country and across the world, and their effects
vary according to the type of physical environment (city centre, suburban, open
moorland, etc.) and the height of the building. Wind loads can act in any plan direction and their intensity can vary continually.

Structural engineers needs to assess the effects of all these loads on any building that
they design, including the two very different structures shown in Figs 5.1 and 5.2.

Lateral (or horizontal) loads other than wind loads include those due to earth pressure (on retaining walls, for example) and water pressure (on the side walls of water
tanks). Other loads may include those due to earthquakes or subsidence.

Why differentiate between these types of loading?

Because the various loads described above are different in nature, we have to deal with
them in different ways when we undertake structural design. For example, the total
dead load in a given building remains constant unless building alterations are carried

Fig. 5.2   Scottish Parliament Building, Edinburgh.
For this unusual – some might say quirky – structure, there were nevertheless no particular structural problems involved in this building's design or construction, but the Scottish Parliament Building was completed three years late and, at £414 million, cost almost ten times the original estimate.

out, but live load can vary on an hour-by-hour – or even minute by minute – basis. We will revisit this when we consider the basics of structural design in Chapter 22.

Nature of load

As well as considering the different types of loading we have to consider the nature of loads. This could be one of three types:

1) point load
2) uniformly distributed load
3) uniformly varying load

Let's consider each of these in turn (see Fig. 5.3).

Point load

This is a load that acts at a single point. It is sometimes called a ***concentrated load***. An example would be a column supported on a beam. As the contact area of the column on the beam would be small, the load is assumed to be concentrated at a point. Point loads are expressed in units of kN and are represented by a large arrow in the direction that the load or force acts, as shown in Fig. 5.3a.

Uniformly distributed load

A uniformly distributed load (often abbreviated to as UDL) is a load that is evenly spread along a length or across an area. For example, the loads supported by a typical beam – the beam's own weight, the weight of the floor slab it's supporting and the live load supported by the floor slab – are consistent all the way along the beam. UDLs along a beam (or any other element that is linear in nature) are expressed in units of kN/m. Similarly, the loads supported by a slab will be consistent across the slab and because a slab has area rather than linear length, UDLs on a slab are expressed in units of kN/m². There are at least two different symbols used for UDL, as shown in Fig. 5.3b.

Uniformly varying load

A uniformly varying load is a load that is distributed along the length of a linear element such as a beam, but instead of the load being evenly spread (as with a UDL) it varies in a linear fashion. A common example of this is a retaining wall. A retaining wall is designed to hold back earth, which exerts a horizontal force on the back of the retaining wall. The horizontal force on the retaining wall becomes greater the further down the wall you go. Thus the force will be zero at the top of the retaining wall but will increase linearly to a maximum value at the bottom of the wall – see Fig. 5.3c.

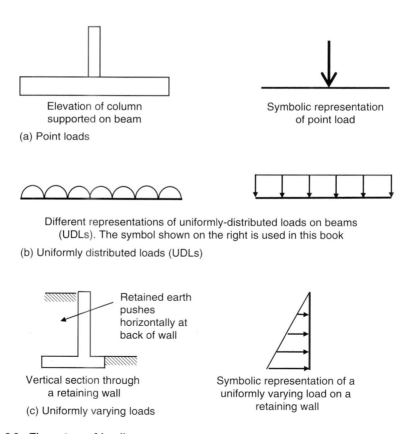

Elevation of column
supported on beam

(a) Point loads

Symbolic representation
of point load

Different representations of uniformly-distributed loads on beams
(UDLs). The symbol shown on the right is used in this book

(b) Uniformly distributed loads (UDLs)

Retained earth
pushes
horizontally at
back of wall

Vertical section through
a retaining wall

(c) Uniformly varying loads

Symbolic representation of a
uniformly varying load on a
retaining wall

Fig. 5.3 **The nature of loading.**

Load paths

It is important to be able to identify the ***paths*** that loads take through a building. As an example, we will consider a typical steel-framed structure, which comprises vertical columns arranged on a grid pattern, as shown in Fig. 5.4a. At each level of the building the columns will support beams, which span between the columns. Each beam may support secondary beams, which will span between the main (primary) beams. The beams will support floor slabs, usually of reinforced concrete or a steel/concrete composite construction. The floor slabs support their own weight and the live loads on them. Figure 5.4b shows a typical part of the structural floor plan. A, C, D and F are columns. The lines AC, AF, DF and CD represent primary beams and line BE represents a secondary beam. A concrete floor slab spans between beams AF and BE. Another concrete floor slab spans between beams BE and CD.

Assume a heavy piece of equipment is located at point G. Clearly, it is supported by the concrete floor slab beneath it. The concrete floor slab, in turn, is supported by beams AF and BE, so it follows that the equipment is supported by those beams as well. As point G is closer to BE than AF, it can be deduced that beam BE takes a greater share of the equipment load than does beam AF.

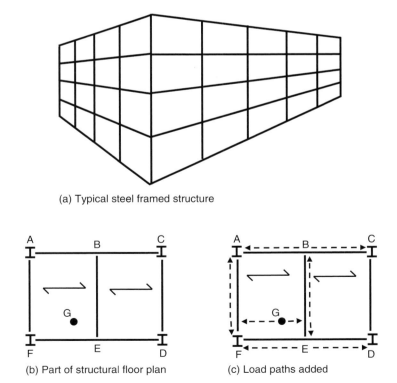

(a) Typical steel framed structure

(b) Part of structural floor plan   (c) Load paths added

Fig. 5.4 **Load paths in a structure.**

Beam BE in turn is supported by primary beams AC and DF. As point G is closer to DF than AC, beam DF will support a greater share of the equipment load than beam AC. (We'll find out how to calculate these 'shares' when we look at **reactions** later in the book.)

Columns D and F support the two ends of beam FD. As beam BE sits exactly half way along DF, it will inflict a point load at the midpoint of DF, which will be shared equally between the two supporting columns D and F. Additionally, column F will support a portion of the equipment load transmitted via beam AF. Similarly, columns A and C will also take a share of the equipment load, via beam AC. Additionally, column A will support a portion of the equipment load transmitted via beam AF.

The broken arrows in Fig. 5.4c indicate the paths taken through the structure by the equipment load at point G. The columns will transmit the equipment loads – along with all the other loads in the structure, of course – down to the foundations, which will need to be strong enough to safely transmit the loads into the ground below.

To simplify the explanation, we've considered only the load paths due to one particular load. Of course, there are many loads in a building and all of them need to be considered in structural analysis and design. We will look at some examples of this in Chapter 24.

It's one thing for a designer to impose a weight limit on a bridge; it's quite another thing to ensure that the weight limit is actually observed.

A marine leisure complex was being constructed on a small tropical island adjacent to a glamorous holiday location. A concrete bridge was designed to link the island with the mainland. As the bridge would be carrying only 'trams' (the sort of trolley-like vehicles common in theme parks) which would transport visitors to the attraction from the car park on the mainland, the bridge was designed for a vehicle weighing 4 tonnes.

During construction of the leisure complex itself it was observed that 25-ton fully laden concrete wagons were crossing the bridge. There were no signs indicating a weight limit and there were no physical barriers to stop anyone crossing. Fortunately, the bridge was clearly grossly over-designed and did not fail – or even show any signs of distress – under these loads.

6 Equilibrium – a balanced approach

Introduction

The eminent scientist Isaac Newton (1642–1727) is perhaps best known for his three laws of motion. If you have studied physics you will have come across these before. One of them gives the *force = mass × acceleration* formula mentioned in Chapter 4.

In this chapter we are concerned with Newton's Third Law of Motion, which essentially states:

'*For every action there is an equal and opposite reaction.*'

This means that if an object is stationary – as a building, or any part of it, usually is – then any force on it must be opposed by another force, equal in magnitude but opposite in direction. In other words, a condition of ***equilibrium*** will be established. See Fig. 6.1 for an example.

Vertical equilibrium

Vertical equilibrium dictates that:

Total force upwards = Total force downwards

For example, if a man weighing about 16 stone stands on the floor of a room, the downward force into the floor due to the weight of his body is 1 kN. Assuming that the floor is stationary, it must be pushing up with an upward force of 1 kN.

Let's consider what would happen if the floor did not react with the same upward force as the downward force encountered. If the floor could, for some reason, muster an upward force of only, say, 0.5 kN in response to the man's downward force of 1 kN, the floor would not be capable of supporting the man's weight. The floor would break and the man would fall through it. On the other hand, if the floor were to react to the man's weight by supplying an upward force of, say, 2 kN, the man would go flying through the air like a human cannonball.

In each of the above two cases, we can see that if the upward and downward forces don't balance, movement occurs (either upward or downward). If neither of these things is occurring (in other words, the man is neither falling through the floor nor shooting through the air), we can conclude that, because everything is stationary, the forces are balanced and vertical equilibrium is observed.

Basic Structures, Second Edition. Philip Garrison.
© 2011 John Wiley & Sons, Ltd. Published 2011 by John Wiley & Sons, Ltd.

Fig. 6.1   Steel arch bridge, Ilkley, West Yorkshire.
The nature of the forces within arches leads to horizontal outward forces being generated at the ends of the arch. For equilibrium to occur, these outward forces must be opposed by inward (i.e. opposite) forces. These inward forces might be in the form of the reaction of a solid abutment to the bridge. Alternatively, as with this bridge, the road deck acts as a horizontal tension member which ties the two ends of the arch to each other, thus catering for the outward thrusting forces.

Horizontal equilibrium

This tells us that:

Total force to the left = Total force to the right

Example 6.1: 'Tug of war'

A tug of war is a physical competition involving two teams and a very long piece of rope. The two teams normally comprise equal numbers of contestants. Each team distributes itself along one end of the rope, as illustrated in Fig. 6.2. The team at the left end of the rope is using all its strength to pull the rope (and the opposing team) to the left. Similarly, the team at the right end of the rope is using all its strength to pull the rope to the right. If there is a river separating the two teams, the stronger team will eventually win by pulling the opposing team into the river.

A marker flag is fixed to the midpoint of the rope. Suppose you are an adjudicator, watching the competition's progress from a distance. You will be watching the marker's position. If the flag starts to move to the left, you will interpret this as meaning that the left-hand team is winning. In other words, the force to the left is greater than the force to the right, and movement occurs because the two forces are unbalanced. Similarly, if the flag starts moving to the right, this would indicate that the right-hand team is winning – because the force to the right is greater than the force to the left. Again, the two forces are unbalanced, causing movement.

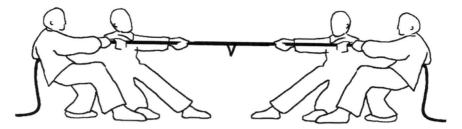

Fig. 6.2   Tug of war.

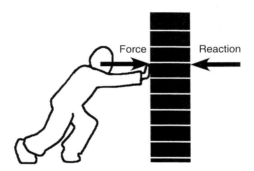

Fig. 6.3   Pushing against a wall.

However, if the flag doesn't move at all but stays in exactly the same position no matter how hard the two teams strain and pull, you would deduce that the two teams are evenly matched and neither is winning. In this case, the flag doesn't move because the force to the left is exactly the same as the force to the right. In other words, if the marker on the rope in a tug of war – or any other object – is stationary, then the force to the left and the force to the right are the same. So we have horizontal equilibrium.

Example 6.2: The leaner

If you lean against a wall, as shown in Fig. 6.3, your body is applying a horizontal force to the wall – to the right in the case shown in Fig. 6.3. The wall will react by providing a force (or reaction) to the left, equal in magnitude to the force applied.

If, for some reason, the wall is not able to provide an equal and opposite horizontal reaction, it means that either the wall is not strong enough or it's not fixed properly to the floor. In either case the wall will yield and movement will occur.

What we've learned about horizontal and vertical equilibrium is summarised in Fig. 6.4.

The application of equilibrium

As buildings are usually stationary, that tells us that the forces on a building – or any part of it – must be in equilibrium.

Consider the beam shown in Fig. 6.5. The beam is supported on columns at each of its ends and supports vertical loads $F1$, $F2$ and $F3$ at various points along its length. Where there are downward forces, there must be opposing upward forces, or reactions. Let's call the reaction at the left-hand end of the beam $R1$. The reaction at the right-hand end of the beam we will call $R2$. Using our knowledge of vertical equilibrium we can say:

Total force up = Total force down

Total force up = total force down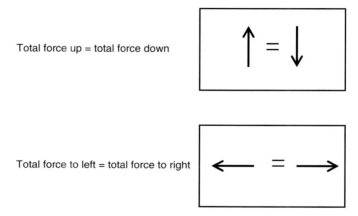

Total force to left = total force to right

Fig. 6.4   Equilibrium.

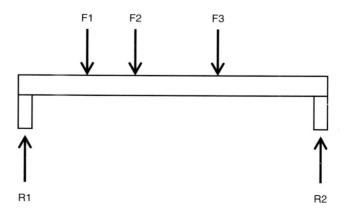

Fig. 6.5   Application of vertical equilibrium.

Fig. 6.6   Tall stone buildings, Honfleur, France.
These historic stone buildings are many storeys high and must therefore apply
quite a large downward forces to the ground underneath, and for its part, the ground
must be strong enough to provide an upward force equal to this downward force –
if it were not capable of doing so, the buildings would have long ago sunk into
the ground.

So:

$$R1 + R2 = F1 + F2 + F3$$

Now if $F1 = 5\,kN$, $F2 = 10\,kN$ and $F3 = 15\,kN$, then:

$$R1 + R2 = 5 + 10 + 15\,kN$$

So

$$R1 + R2 = 30\,kN$$

It would be useful to calculate $R1$ and $R2$, as they represent the forces in the supporting columns. But the equation above doesn't tell us what $R1$ is and it doesn't tell us what $R2$ is; it merely tells us that the sum of the two is $30\,kN$. In order to evaluate each of $R1$ and $R2$, we need to know more. We will continue this theme in Chapter 9. In the meantime, see Figure 6.6 for a practical application of equilibrium.

The late Yorkshire veterinary surgeon James Herriot describes an encounter with a bull in a confined space in his book *Vet in Harness* (Michael Joseph, 1974). The bull, which had just received an injection from Mr Herriot, decided to lean against the vet, thus sandwiching him between its body and a wooden partition. As we know from Chapter 3, this action would have put Mr Herriot's body into compression, held in place by opposing reactions from the bull and the wooden partition. As he puts it: 'I was having the life crushed out of me. Pop-eyed, groaning, scarcely able to breathe, I struggled with everything I had, but I couldn't move an inch. ... I was certain my internal organs were being steadily ground to pulp and as I thrashed around in complete panic the huge animal leaned even more heavily.

'I don't like to think what would have happened if the wood behind me had not been old and rotten, but just as I felt my senses leaving me there was a cracking and splintering and I fell through into the next stall.'

So suddenly – and fortunately for Mr Herriot – the force from the bull overcame the wooden partition's strength and it collapsed. It was no longer able to provide a reaction to keep Mr Herriot in compression and thus probably saved his life.

What you should remember from this chapter

If any object (for example, a building or part of one) is stationary then it is in equilibrium. This means that the forces on it must balance, as follows:

- Total force upwards = Total force downwards
- Total force to the left = Total force to the right

7 More about forces: resultants and components

Introduction

In previous chapters you have learned what a force is. In this chapter we will look at how forces – individually or in groups – may be handled. You will learn how to combine forces into *resultants* and how to 'split' forces into **components**.

Let's start by considering an analogy.

The Underground analogy

Imagine that you are in London and are planning a journey on the Underground railway system there. You are at Green Park station and want to travel to Oxford Circus. You consult the diagram of lines and stations displayed at the station entrance, a representation of the relevant part of which is shown in Fig. 7.1. You work out that the quickest way to reach Oxford Circus from Green Park is to travel directly there on the Victoria Line. Oxford Circus is only one station from Green Park.

However, as you enter the station, you pass a blackboard on which has been written: 'Victoria Line Closed due to Technical Difficulties'. Clearly, this news means that you must change your travel plans. Assuming that you don't now decide to walk or take a bus or taxi, there are two options available to you if you wish to reach Oxford Circus as quickly as possible:

1) Take the Jubilee Line northwards to the next station, Bond Street, where you can change onto the eastbound Central Line and travel to the next station, Oxford Circus.
2) Take the Piccadilly Line eastwards to the next station, Piccadilly Circus, where you can change onto the northbound Bakerloo Line and travel to the next station, Oxford Circus.

Clearly one of these two options will be quicker than the other, depending on frequency of trains and the ease of transferring between platforms at the interchange station. Although it is difficult to predict which option would deliver you to Oxford Circus more quickly, we can say with confidence that either route will take you – eventually – to Oxford Circus.

Basic Structures, Second Edition. Philip Garrison.
© 2011 John Wiley & Sons, Ltd. Published 2011 by John Wiley & Sons, Ltd.

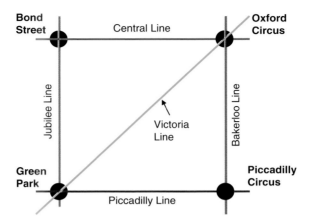

Fig. 7.1   Extract from London Underground network.

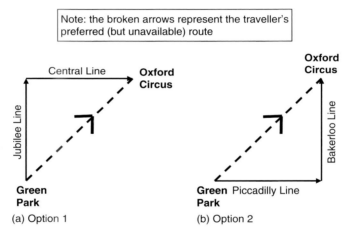

Fig. 7.2   Green Park to Oxford Circus route options.

If we represent a journey by an arrow in the direction of the journey – with the length of the arrow representing the length of the journey – our two route options can be illustrated by the two diagrams in Fig. 7.2. In each case, the desired direct route (on the temporarily unavailable Victoria Line) has been indicated by a broken arrow. As expected, each indirect route is longer (in distance) than the direct route between Green Park and Oxford Circus stations, but the end result is the same: in each case, you end up at Oxford Circus station.

Whichever option you choose, your starting point is Green Park station and your finishing point is Oxford Circus station.

We will return to this analogy later in the chapter.

Resolution of forces

We encountered the concept of a force in Chapter 4. As stated there, a force is an influence or action on a body that causes – or attempts to cause – movement. Forces can act in any direction, but the direction in which a given force acts is important. You will know from your studies of mathematics that something that has both magnitude and direction is called a *vector* quantity. As force has both magnitude and direction, force is an example of a vector quantity.

To define a given force fully, we need to state its:

- magnitude (for example, 50 kN)
- direction, or line of action (for example, vertical)
- point of application (for example, 2 metres from the left-hand end of a beam).

What happens when several forces act at the same point?

Clearly it is possible that several forces may act at the same point. These forces may all be different in magnitude and acting in different directions. It would be convenient if we could simplify these forces in such a way that they are represented by just one force, acting in a certain direction. This one force is called the *resultant* force.

The 'Donald and Tristan' analogy

Consider a trolley standing in the middle of a large room with a highly polished wooden floor. The trolley is a piece of furniture with wheels or castors that make it easy to push it in any direction – rather like the dessert trolleys used in expensive restaurants. The room is otherwise empty. Donald enters the room and starts to push the trolley in an easterly direction. The trolley moves eastwards, as shown in Fig. 7.3a. At that moment, Donald's friend Tristan enters the room and starts to push the trolley northwards, while Donald continues to try to push the trolley eastwards.

As you would expect, under the influence of the two friends pushing the trolley in different directions, the trolley now moves off in a generally north-easterly direction. Figure 7.3b indicates this activity, as viewed from above (a plan view), with the broken arrow representing the movement of the trolley. But in what direction, exactly, would the trolley move?

Well, it depends on the relative effort that Donald and Tristan put into the exercise. If Donald puts a lot of energy into his easterly push, while Tristan makes a puny attempt at his northbound shove, the trolley will move off in the direction shown by the broken arrow in Fig. 7.3c. (The length of the arrows represents the magnitude of the forces.) On the other hand, if Tristan exerts himself fully with his northerly push and Donald can't be bothered to put much effort into his eastbound push, the trolley will move off in the direction shown by the broken arrow in Fig. 7.3d. (Again, the length of the arrows represents the magnitude of the forces.) In each case there are two forces involved: one from Donald, the other from Tristan. As we have seen, these two forces are of different magnitudes and act in different directions.

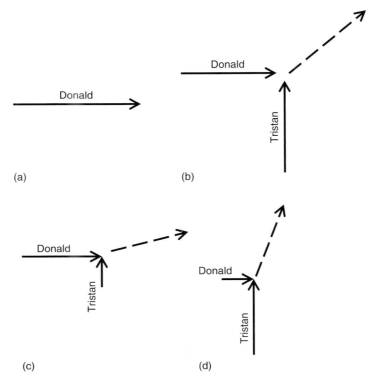

Fig. 7.3 Resultants of forces.

In each of Figs 7.3b–d the broken arrow represents the ***resultant*** force. In each case, the magnitude and direction of this resultant force represents the combined effect of Donald and Tristan's pushing. If we knew the magnitude of the force (that is, how many kN) that each of the two men was putting into the exercise, we could calculate the ***magnitude*** – and exact ***direction*** – of the resultant force.

Of course, we may have more than two forces. Donald and Tristan's mutual friend Tarquin may enter the room and start pushing the trolley in a different direction while Donald and Tristan are exerting themselves. The trolley would move off in a different direction. Again, the direction of movement of the trolley represents the direction of the resultant force.

Resultants of forces

The resultant force (let's call it R) is the single force that would have the same effect on an object as a system of two or more forces. Resultants can be calculated by simple trigonometry or by graphical methods. Example 7.1 shows how trigonometry may be used.

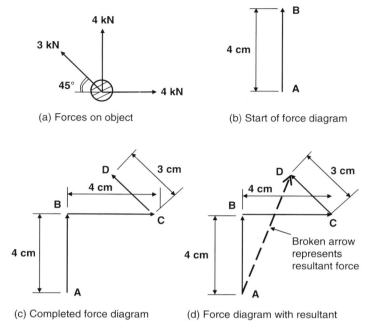

(a) Forces on object

(b) Start of force diagram

(c) Completed force diagram

(d) Force diagram with resultant

Fig. 7.4 Object subjected to three forces.

Example 7.1

An object is subjected to three forces of different magnitudes, acting in different directions, as shown in Fig. 7.4a. Using a graphical approach, determine the magnitude and direction of the resultant force.

We can consider the forces in any order. We will consider the vertical 4 kN force first. Using graph paper and choosing a suitable scale (1 cm = 1 kN, in this case), we will start from a point A. The vertical force can be represented by a line vertically upwards from point A, 4 cm long (to represent 4 kN). The point we arrive at will be called point B (see Fig. 7.4b). Next, let's consider the horizontal 4 kN force, which acts to the right. Starting from point B, draw a horizontal line 4 cm long (going to the right), representing the horizontal force of 4 kN. The point we arrive at will be point C.

Finally, let's consider the 3 kN force, which acts diagonally upwards to the left at an angle of 45°. Starting from point C, this force will be represented by a line 3 cm long (representing 3 kN) in the appropriate direction. The point we arrive at will be point D (see Fig. 7.4c). Next, draw a straight line connecting points A and D. This line represents the resultant force. Measuring off the diagram (Fig. 7.4d) it can be found that the line is 6.41 cm long at an angle to the horizontal of 72.9°.

Therefore the resultant force is 6.41 kN, acting at an angle of 72.9° to the horizontal (upwards and to the right). This is the single force that would have the same effect as the original three forces acting together.

This problem could alternatively have been approached mathematically, using Pythagoras' theorem and basic trigonometry, which are summarised in Appendix 3. The mathematical solution to this problem is shown in Fig. 7.5.

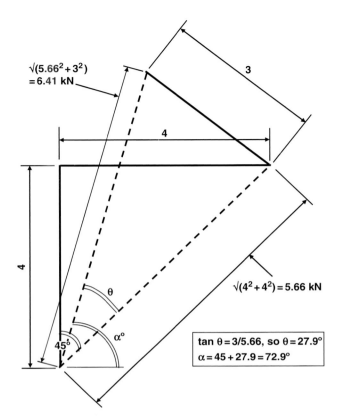

$\sqrt{(5.66^2 + 3^2)}$
= 6.41 kN

3

4

4

θ

$\alpha°$

$\sqrt{(4^2 + 4^2)} = 5.66$ kN

$45°$

tan θ = 3/5.66, so θ = 27.9°
α = 45 + 27.9 = 72.9°

Fig. 7.5   Mathematical solution to resultants example.

Further examples

Each of the examples shown in Fig. 7.6 comprises two forces at right angles to each other. In each case, the task is to find the magnitude and direction of the resultant force. This can be calculated either mathematically or graphically.

To determine the resultants mathematically you will need to reacquaint yourself with the basic mathematics associated with a right-angled triangle, namely Pythagoras' theorem, and the definitions of the trigonometrical functions known as sine, cosine and tangent. Appendix 3 gives you a quick refresher on these. To determine the resultants graphically, you need to represent the forces by lines on graph paper whose lengths are proportional to the magnitudes of the forces. The lines need to be orientated in the same directions as the corresponding forces.

Whichever way you determine the resultant, it doesn't matter which order you consider the forces in; you will still get the same answer – this was the point made by the 'Underground analogy' earlier in the chapter when we saw that there is more than one route from Green Park to Oxford Circus. However, you **must** consider the forces in the 'nose-to-tail' manner adopted in the example above (and shown in Fig. 7.4c), otherwise your answer for the direction will be wrong. In my experience, by far the most common mistake students make when dealing with this sort of problem lies in

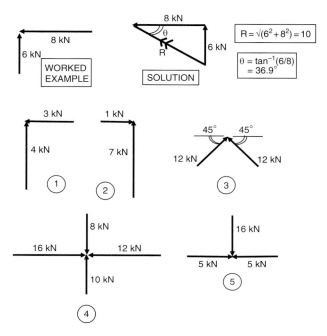

Fig. 7.6 Components examples.

their not redrawing the forces 'nose to tail'. So make sure that the 'nose' (that is, the arrowed end) of each force is laid adjacent to the 'tail' (the non-arrowed end) of the next force because 'nose to nose' or 'tail to tail' will give the wrong answer.

Study the worked example given at the top of Fig. 7.6. Note how it has been solved, first of all by expressing the forces in a nose-to-tail manner, then by calculating the magnitude of the resultant force using Pythagoras' theorem and its direction using trigonometry.

Now attempt the other five examples given in Fig. 7.6. Figure 7.7 shows the solutions to the examples shown in Fig. 7.6.

If you got the direction of the first two examples wrong, then you haven't been expressing the forces in the nose-to-tail manner required; in each case, the problem has to be reconstructed in the manner shown in Fig. 7.7. If you got the first two examples right but came to a dead halt when you reached example number 3, then your mathematical knowledge of right-angled triangles is probably fine but you've lost sight of what resultants actually are. Remember: to obtain the resultant of two or more forces you express the forces (in any order) in a nose-to-tail fashion, then you draw a line linking the tail of the first force with the nose of the final force. In the case of example 3, this resultant force turns out to be vertically upwards.

You should have realised that examples 4 and 5 can be simplified. For instance, in example 4, the 16 kN force to the right is partially cancelled by the 12 kN force to the left, to give an overall force to the right of 4 kN (i.e. 16 − 12). Similarly, the upward force will be 2 kN (i.e. 10 − 8).

You will find further examples at the end of this chapter.

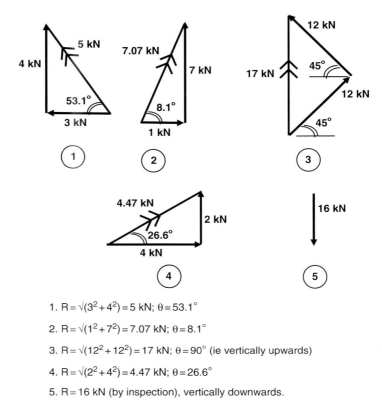

1. $R = \sqrt{(3^2 + 4^2)} = 5$ kN; $\theta = 53.1°$

2. $R = \sqrt{(1^2 + 7^2)} = 7.07$ kN; $\theta = 8.1°$

3. $R = \sqrt{(12^2 + 12^2)} = 17$ kN; $\theta = 90°$ (ie vertically upwards)

4. $R = \sqrt{(2^2 + 4^2)} = 4.47$ kN; $\theta = 26.6°$

5. $R = 16$ kN (by inspection), vertically downwards.

Fig. 7.7   Components: solutions to examples.

Components of forces

Earlier in this chapter we looked at how we could express a number of different forces, acting together, at the same point, as a single force – the resultant. Now we are going to invert the process by taking a single force and breaking it down into two forces which, taken together, have the same effect as the original single force.

These two forces are called ***components.*** In the same way as a television set contains many electronic components, all of which must be present for the television to work, so must both of our force components be present to correctly represent the original force. A component is the replacement of an original force with two forces at right angles to each other (usually one horizontal and one vertical).

It can be shown that, for any force F at an angle θ to the horizontal, the horizontal component is always $F.\cos \theta$ and the vertical component is always $F.\sin \theta$ (see Fig. 7.8). As a mnemonic, think of 'sign up' to represent sine being the vertical force.

For each of the three examples given in Fig. 7.9, calculate the magnitude and direction of the two components (one horizontal, the other vertical) of the force given.

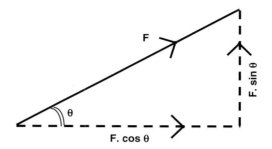

Fig. 7.8 Components of forces – general case.

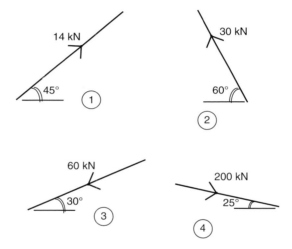

Fig. 7.9 Components of forces – examples.

In each case, make sure that you correctly identify whether the horizontal force is to the left or to the right, and whether the vertical force is upwards or downwards – such things are important! Check your answers with the following (where H = horizontal component and V = vertical component):

1) $H = 14.\cos 45° = 9.9\,\text{kN} \rightarrow$, $V = 14.\sin 45° = 9.9\,\text{kN} \uparrow$.
2) $H = 30.\cos 60° = 15\,\text{kN} \leftarrow$, $V = 30.\sin 60° = 26\,\text{kN} \uparrow$.
3) $H = 60.\cos 30° = 52\,\text{kN} \leftarrow$, $V = 60.\sin 30° = 30\,\text{kN} \downarrow$.
4) $H = 20.\cos 25° = 181\,\text{kN} \rightarrow$, $V = 200.\sin 25° = 84.5\,\text{kN} \downarrow$.

We shall see later in this book how useful it is to be able to replace a force acting at an angle by two forces: one horizontal and the other vertical.
    For a real-life application of this, see Fig. 7.10.

Fig. 7.10   Cable stayed walkway, Dublin Airport.
This covered walkway is suspended from cables supported by masts; the designers will
have needed to calculate the vertical and horizontal components of the forces in these
inclined cables.

What you should remember from this chapter

- The resultant of a number of forces acting at a point is the single force which has the same effect as the original forces acting together.
- Any force can be split up into two components which, acting together, have the same effect as the original single force. The two components are at right angles to each other and are usually taken as horizontal and vertical respectively.

Tutorial examples

Find the resultant for each of the multi-force examples shown in Fig. 7.11 and split each of the one-force examples into components.

Further tutorial examples

1. Calculate the magnitude and direction of the resultant of the system of forces shown in Fig. 7.12a.
2. If the four forces shown in Fig. 7.12b are in equilibrium (and therefore the resultant force is zero) calculate force $F1$ and angle θ.

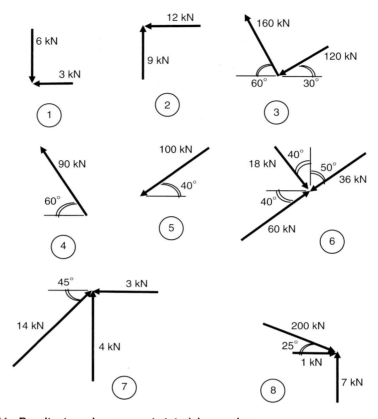

Fig. 7.11   **Resultants and components tutorial examples.**

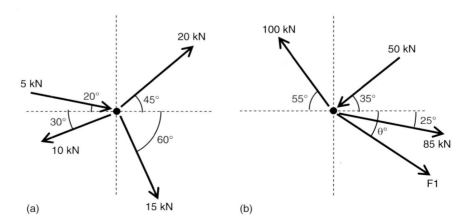

Fig. 7.12   **Further tutorial examples.**

8 Moments

Designed by British architect Sir Norman Foster, the modern glass dome shown in Fig. 8.1 is the historic German parliament building's crowning glory.

Introduction

When you use a spanner to tighten or loosen a nut, you are applying a *moment*. A moment is a turning effect and is related to the concept of leverage. If you use a screwdriver to prise open the lid on a can of paint, or a bottle opener to open a bottle of beer, or a crowbar to lift a manhole cover, you are applying leverage and hence you are applying a moment.

What is a moment?

A moment is a turning effect. A moment always acts about a given point and is either clockwise or anticlockwise in nature. The moment about a point A caused by a particular force F is defined as the force F multiplied by the perpendicular distance from the force's line of action to the point.

Units of moment are kN.m or N.mm. Note: This follows because a moment is a force multiplied by a distance, therefore its units are the units of force (kN or N) multiplied by distance (m or mm) – hence kN.m or N.mm. The units of moment are *never* kN or kN/m.

In both the cases illustrated in Fig. 8.2, if M is the moment about point A, then $M = F.x$.

Practical examples of moments

Seesaw

A see-saw is a piece of equipment often found in children's playgrounds. It comprises a long plank of wood with a seat at each end. The plank of wood is supported at its centre point. The support is pivoted, so is free to rotate. One child sits on the seat at

Basic Structures, Second Edition. Philip Garrison.
© 2011 John Wiley & Sons, Ltd. Published 2011 by John Wiley & Sons, Ltd.

Fig. 8.1   **Reichstag dome, Berlin.**

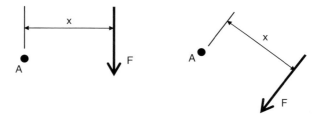

Fig. 8.2   **Moments illustrated.**

Fig. 8.3   **Seesaws.**

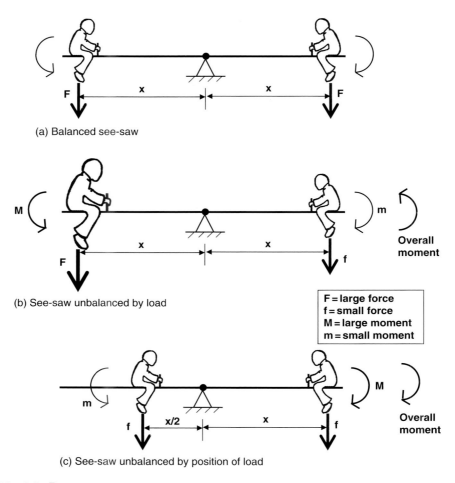

(a) Balanced see-saw

(b) See-saw unbalanced by load

F = large force
f = small force
M = large moment
m = small moment

(c) See-saw unbalanced by position of load

Fig. 8.4   Forces on a see-saw

each end of the see-saw and uses the pivoting characteristic of the see-saw to move up and down. A series of modern see-saws is shown in Fig. 8.3.

Imagine two young children sitting at opposite ends of a see-saw, as shown in Fig. 8.4a. If the two children are of equal weight and sitting at equal distances from the see-saw's pivot point, there will be no movement because the clockwise moment about the pivot due to the child at the right-hand end $(F \times x)$ is equal to the anticlockwise moment about the pivot point due to the child at the left-hand end $(F \times x)$. Therefore the two moments cancel each other out.

If the child at the left-hand end was replaced by an adult or a much larger child, as shown in Fig. 8.4b, the child at the right-hand end would move rapidly upwards. This is because the (anticlockwise) moment due to the larger person at the left-hand end (large force × distance) is greater than the (clockwise) moment due to the small child at the right-hand end (small force × distance). The overall moment is thus anticlockwise, causing upward movement of the small child.

Let's return to the original situation, with two young children at opposite ends of the see-saw. But suppose now that the left-hand child moved closer to the pivot point, as shown in Fig. 8.4c. As a result, the right-hand child would move downwards. This is because the anticlockwise moment due to the left-hand child (force × small distance) is smaller than the clockwise moment due to the right-hand child (force × large distance). The overall movement is thus clockwise, causing downward movement of the right-hand child.

Spanners, nuts and bolts

The reader will know from experience that it is much easier to undo a seized-up nut or bolt if a long spanner is used rather than a short spanner. This is because, although the force used may be the same, the 'lever arm' distance is longer, thus causing a greater turning effect or moment to be applied. Practical problems using 'leverage' also illustrate this principle, such as the examples of prising open paint cans, beer bottles and manhole covers already mentioned.

Numerical problems involving moments

It can be seen from the above that it is important to distinguish between clockwise and anticlockwise moments. After all, turning a spanner clockwise (tightening a nut) has a very different effect from turning the spanner anticlockwise (loosening a nut). In this book:

- clockwise moments are regarded as positive (+)
- anticlockwise moments are regarded as negative (−).

It is, of course, quite possible for a given pivot point to experience several moments simultaneously, some of which may be clockwise (+) while others may be anticlockwise (−). In these cases, moments must be added algebraically to obtain a total (net) moment.

Some simple worked examples of moment calculation

In each of the following examples, involving simple beams, we are going to calculate the net moment about point A (remember, clockwise is +, anticlockwise is −).

Example 8.1 (see Fig. 8.5a)
By inspection, the 4 kN force is trying to turn clockwise about A, so the moment will be positive (+). The 2 metre distance is measured horizontally from the (vertical) line of action of the 4 kN force; in other words, the distance given is measured perpendicular (i.e. at right angles or 90°) to the line of action of the force, as required.

Remember, a moment is a force multiplied by a distance. If we use the symbol M to represent moment, then in this case:

$$M = +(4\,kN \times 2\,m) = +8\,kN.m$$

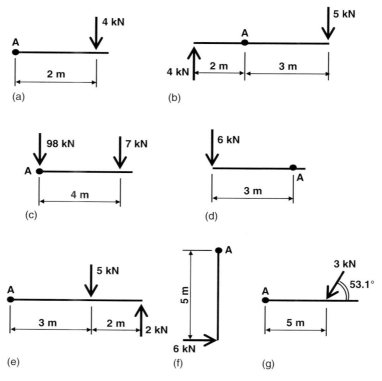

Fig. 8.5   Moment worked examples.

Example 8.2 (see Fig. 8.5b)

This time there are two forces, supplying two moments. A common mistake with this example is to assume that since the two forces are in opposite directions (i.e. one upwards, one downwards), the moments must also oppose each other. In fact, a closer inspection will reveal that the moments about A generated by the two forces are both clockwise (+). So the moment about A for each force is calculated, and the two added together, as follows:

$$M = +(5\,\text{kN} \times 3\,\text{m}) + (4\,\text{kN} \times 2\,\text{m})$$
$$= +15\,\text{kN.m} + 8\,\text{kN.m}$$
$$= +23\,\text{kN.m}$$

(If you attempted this example and obtained an answer of +7 kN.m, you fell into the trap mentioned above!)

Example 8.3 (see Fig. 8.5c)

Once again there are two forces, supplying two moments. The 7 kN force clearly gives rise to a clockwise moment about A. The 98 kN force, however, passes straight through the pivot point A; in other words, its line of action is zero distance from A. Since a moment is always a force multiplied by a distance, if the distance is zero then it follows that the moment must be zero (since multiplying any number by zero gives a product of zero). So, in this example:

$$M = +(7\,kN \times 4\,m) + (98\,kN \times 0\,m)$$
$$= +28\,kN.m + 0\,kN.m$$
$$= +28\,kN.m$$

The lesson to be learned from this example is: if a force passes through a certain point, then the moment of that force about that point is zero.

Example 8.4
In Fig. 8.5d, the 6 kN force is turning anticlockwise about A, so the resulting moment will be negative ($-$).

$$M = -(6\,kN \times 3\,m) = -18\,kN.m$$

Example 8.5
In Fig. 8.5e, the 5 kN force is trying to turn clockwise about A, therefore will give rise to a clockwise (+) moment. By contrast, the 2 kN force is trying to turn anticlockwise about A, so it will produce an anticlockwise ($-$) moment.

$$M = +(5\,kN \times 3\,m) - (2\,kN \times 5\,m)$$
$$= +15\,kN.m - 10\,kN.m$$
$$= +5\,kN.m$$

Example 8.6
In Fig. 8.5f, not all forces are vertical! But the same rules apply.

$$M = -(6\,kN \times 5\,m) = -30\,kN.m$$

Example 8.7
The slightly harder example in Fig. 8.5e will confuse readers who haven't yet grasped the fact that a moment is a force multiplied by a *perpendicular* (or 'lever arm') distance. There are two ways of solving this problem – see Fig. 8.6.

First, you can use trigonometry to find the perpendicular distance. Figure 8.6a will remind you of the definitions of sines, cosines and tangents in terms of the lengths of the sides of a right-angled triangle. Applying this to the current problem, we find from Fig. 8.6b that the perpendicular distance, x, in this case is 4 metres. So $M = +(3\,kN \times 4\,m) = +12\,kN.m$.

Second, we can solve it by resolving the 3 kN force into vertical and horizontal components. In Chapter 7 we learned that any force can be expressed as the resultant of two components, one horizontal and one vertical. For any force F acting at an angle of θ to the horizontal axis, it can be shown that:

- the horizontal component is always $F \times \cos \theta$
- the vertical component is always $F \times \sin \theta$ (as sin acts upwards: 'sign up').

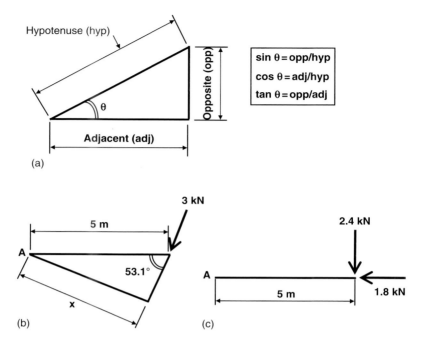

Fig. 8.6   Moments and resolution of forces.

In this problem, the 3 kN force acts at an angle of 53.1° to the horizontal. So its vertical component = 3 × sin 53.1° = 2.4 kN. ↓

And its horizontal component = 3 × cos 53.1° = 1.8 kN. ←

The problem can now be expressed as shown in Fig. 8.6c.

Note that since the 1.8 kN force (extended) passes through point A, the moment of that force about point A will be zero.

$$M = +(2.4\,\text{kN} \times 5\,\text{m}) + (1.8\,\text{kN} \times 0)$$
$$= +12\,\text{kN.m}$$

(Obviously, this is the same answer as that obtained by the first method!)

There are some further examples at the end of the chapter for you to try.

Notes on moment calculations

From the above discussion and examples come the following notes and observations:

- Always consider whether a given moment is clockwise (+) or anticlockwise (−).
- If a force *F* passes *through* a point A, then the moment of force *F* about point A is zero (as illustrated in Example 8.3 above).
- It may be necessary to resolve forces into components in order to calculate moments (as shown in Example 8.6 above).

Moment equilibrium

Imagine that you are doing some mechanical work on a car and you have a spanner fitted onto a particular bolt in the car's engine compartment. You are trying to tighten the nut and hence you are turning it clockwise. In other words, you are applying a clockwise moment. Now imagine that a friend (in the loosest possible sense of the word!) has another spanner fitted to the same nut and is turning it anticlockwise. This has the effect of loosening the nut.

If the anticlockwise moment that your friend is applying is the same as the clockwise moment that you are applying (regardless of the fact that the two spanners might be of different lengths), you can imagine that the two effects would cancel each other out – in other words, the nut would not move. This applies to any object subjected to equal turning moments in opposite directions: the object would not move.

So if the total clockwise moment about a point equals the total anticlockwise moment about the point, no movement can take place. Conversely, if there is no movement (as is usually the case with a building or any part of a building), then clockwise and anticlockwise moments must be balanced. This is the principle of moment equilibrium, which can be used in conjunction with the rules of force equilibrium (discussed earlier) to solve structural problems.

To summarise: if any object (such as a building or any point within a building) is stationary, the net moment at the point will be zero. In other words, clockwise moments about the point will be cancelled out by equal and opposite anticlockwise moments.

Equilibrium revisited

As discussed in Chapter 6, if an object, or a point within a structure, is stationary, we know that forces must balance, as follows (the Σ symbol, a Greek sigma, means 'the sum of'):

$\Sigma V = 0$, i.e. total upward force = total downward force ($\uparrow = \downarrow$)
$\Sigma H = 0$, i.e. total force to the left = total force to the right ($\leftarrow = \rightarrow$)

From our newly acquired knowledge of moments, we can add a third rule of equilibrium:

$\Sigma M = 0$, i.e. total clockwise moment = total anticlockwise moment

We can use these three rules of equilibrium to solve structural problems, specifically the calculation of end reactions, as discussed in the next chapter.

Couples

As you know, the word *couple* means two. In the context of moments, a couple is a system of two forces of equal magnitude, a certain distance apart, acting in opposite directions.

An example of a couple is shown in Fig. 8.7. In this case, two forces of 30 kN act in opposite directions, and their lines of action are 0.8 metres apart. Let's calculate the moment about each of points A–E:

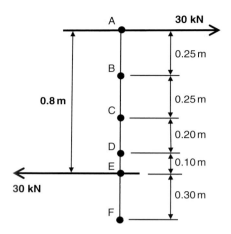

Fig. 8.7   **Example of a couple.**

Moment about point A = (30 kN × 0 m) + (30 kN × 0.8 m) = 24 kN.m
Moment about point B = (30 kN × 0.25 m) + (30 kN × 0.55 m) = 24 kN.m
Moment about point C = (30 kN × 0.5 m) + (30 kN × 0.3 m) = 24 kN.m
Moment about point D = (30 kN × 0.7 m) + (30 kN × 0.1 m) = 24 kN.m
Moment about point E = (30 kN × 0.8 m) + (30 kN × 0 m) = 24 kN.m
Moment about point F = (30 kN × 1.1 m) − (30 kN × 0.3 m) = 24 kN.m

As you can see, it doesn't matter which point you take moments about, the moment will always have the same value for any given couple.

To take a general case: if a couple of forces have a value of *F* kN each, and the forces are *x* metres apart, the moment of that couple of forces about any point will always be *F* multiplied by *x*. In the above example, the moment about any point is 30 kN × 0.8 m = 24 kN.m.

Retaining walls: an illustration of equilibrium

A retaining wall is a wall with a particular structural function: it holds back earth. If the ground level on one side of a wall is different from that on the other, then the wall is acting as a retaining wall. The bigger the difference in level, the greater the earth pressure acting horizontally on the wall, and therefore the stronger the wall has to be to resist this pressure. See Fig. 8.8.

You may have retaining walls in your back garden, particularly if the garden as a whole is sloping. In this situation the height of earth held back is probably small, perhaps less than half a metre, therefore a retaining wall of brick or stone is usually more than adequate for the job.

More significant retaining walls can be found defining cuttings for railways or highways, particularly in cities where land is at a premium and there isn't space for a gently sloping embankment to accommodate the difference in levels. In this situation, the difference in ground level between the two sides of the wall might be several metres, and such retaining walls will usually be of concrete, though they may be faced with brick or stone to make them more attractive.

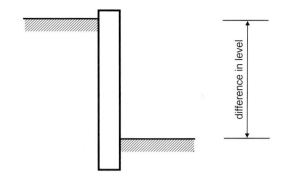

Fig. 8.8 **Retaining walls – the concept.**

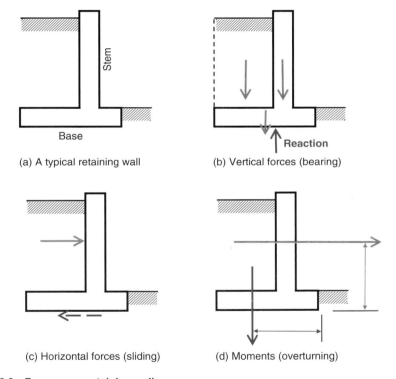

(a) A typical retaining wall

(b) Vertical forces (bearing)

(c) Horizontal forces (sliding)

(d) Moments (overturning)

Fig. 8.9 **Forces on a retaining wall.**

A cross section through a typical concrete retaining wall is shown in Fig. 8.9a. It comprises a horizontal base slab which must be rigidly connected to a vertical wall, or "stem". The stem acts as a vertical cantilever and must be designed accordingly. There are three stability issues with retaining walls:

1. *Bearing*: The weight of the wall (and anything on top of it) must not be so great that it sinks vertically into the ground.

2. *Sliding*: The force from the retained earth must not be so great that the wall, as a whole, slides forward.
3. *Overturning*: The moment from the force on the wall must not be so great that the wall turns over.

Let's consider each of these three effects in turn.

Bearing

As we've seen, for equilibrium the total downward force must equal the total upward force. In other words, the ground must be strong enough to provide a total upward force (or **reaction**, as we'll see in the next chapter) to oppose the downward force due to the weight of the retaining wall – see Fig. 8.9b. If the ground isn't strong enough to provide this upward force, equilibrium will not occur and the wall will sink into the ground.

Sliding

The force of the earth pushing horizontally against the retaining wall (to the right in Fig. 8.9c) must be resisted by an equal force to the left, if the second rule of equilibrium mentioned above (horizontal forces must balance) is to be satisfied. This force to the left must be provided by friction between the base of the wall and the ground beneath. If there isn't enough friction, the frictional force to the left will not be great enough to provide resistance, equilibrium will not occur and the wall will move to the right.

Overturning

The force of the earth pushing against the wall will potentially lead to clockwise rotation of the wall about the front edge of the base (see Fig. 8.9d). As we know from earlier in this chapter, such rotation would be caused by a clockwise moment. If the wall is not to physically rotate, equilibrium will be assured if sufficient anticlockwise moment is provided by the weight of the wall acting downwards.

Consider the retaining wall shown in Fig. 8.10. The force causing the wall to overturn, the weight of the wall and the frictional resistance force have been calculated and are shown on the diagram. The 40 kN vertical force is due to the weight of the concrete forming the stem, and the 30 kN vertical force is due to the weight of the concrete forming the base. The 130 kN vertical force is due to the rectangular 'block' of earth above the base, and the 75 kN horizontal force to the right is the overall effect of the horizontal earth pressure on the wall. We can check equilibrium as follows:

Vertical equilibrium:
Total downward force due to weight of wall and earth above = 130 + 30 + 40 = 200 kN.
Total upward force (reaction) = 200 kN.
Therefore total downward force ↓ = total upward force ↑, so there is vertical equilibrium.

Horizontal equilibrium:
Total rightward force due to horizontal earth pressure = 75 kN.

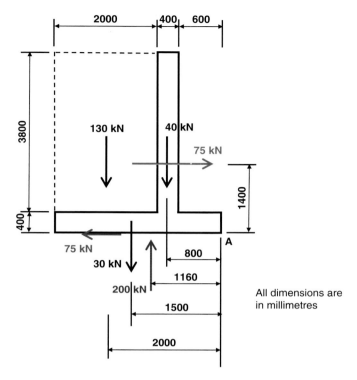

Fig. 8.10   Retaining wall example.

Total leftward force due to friction between underside of base and ground beneath = 75 kN.
Therefore total rightward force → = total leftward force ←, so there is horizontal equilibrium.

Moment equilibrium:
Clockwise moments about point A = (75 kN × 1.4 m) + (200 kN × 1.16 m) = 337 kN.m.
Anticlockwise moments about point A = (130 kN × 2.0 m) + (30 kN × 1.5 m) + (40 kN × 0.8 m) = 337 kN.m.
Therefore clockwise moment = anticlockwise moment, so there is moment equilibrium.

As the three equilibrium equations are all satisfied, the retaining wall is in equilibrium and will remain stationary under the forces and moments to which it is subjected.

A bathplug: an illustration of moments

The photos in Fig. 8.11 show a swivel type bathplug in the closed and open positions respectively. This type of bathplug is connected to the washbasin by a horizontal pin through the diameter of the plug. The plug can be opened or closed by using a finger to turn it about the axis of the pin. A rubber seal prevents leakage.

(a)

(b)

Fig. 8.11   **(a) A swivel bathplug closed (b) A swivel bathplug open.**

When the bath has water in it, the weight of the water on each of the semicircular parts of the plug separated by the pin is the same, and hence the moments due to the water's weight about the pin are the same. These moments are equal in magnitude and opposite in direction, so the plug is in equilibrium and will not attempt to turn. However, pressure from a finger pressing on one side of the plug will introduce an additional moment, so the plug will no longer be in equilibrium and it will open. Note that, despite the (possibly quite large) water pressure on the plug, only a small force will be required to open it.

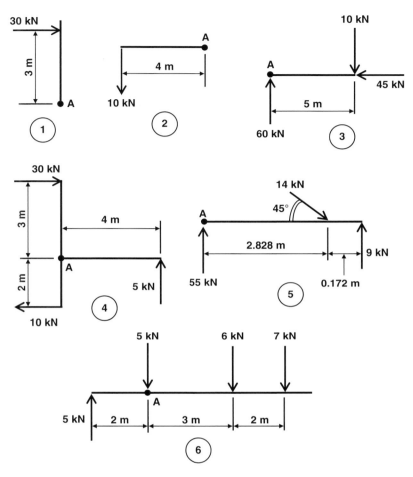

Fig. 8.12 **Moment tutorial examples.**

What you should remember from this chapter

- A moment is one of the most important concepts in structural mechanics.
- A moment is a turning effect, either clockwise or anticlockwise, about a given point.
- If a force passes through the point about which moments are being taken, then the moment of that force about the point concerned is zero. (It is very important to remember this concept as it crops up several times in the solution of problems later in this book.)
- It may be necessary to resolve forces into components in order to calculate moments.

Tutorial examples

In each of the examples shown in Fig. 8.12, calculate the net moment, in kN.m units, about point A.

Tutorial answers

1) $M = +90\,\text{kN.m}$
2) $M = -40\,\text{kN.m}$
3) $M = +50\,\text{kN.m}$
4) $M = +90\,\text{kN.m}$
5) $M = +1\,\text{kN.m}$
6) $M = +63\,\text{kN.m}$

9   Reactions

Note: the support symbols used in the diagrams in this chapter will be explained in Chapter 10.

Introduction

In Chapter 6 we discussed equilibrium. We found out that if a body or object of any sort is stationary, then the forces on it balance, as follows:

Total force upwards = Total force downwards
Total force to the left = Total force to the right

This was summarised in Fig. 6.4.

The concept of a moment, or turning effect, was introduced in Chapter 2 and discussed more fully in Chapter 8. In this chapter we will find out how to use this information to calculate *reactions* – that is, the upward forces that occur at beam supports in response to the forces on the beam.

Moment equilibrium

At the end of Chapter 8 we found that if an object or body is stationary, it doesn't rotate and the total clockwise moment about any point on the object is equal to the total anticlockwise moment about the same point. This is the third rule of equilibrium and we can add this to the first two that we discovered in Chapter 6. The three rules of equilibrium are expressed in Fig. 9.1.

In Figure 9.2 each steel beam imposes a downward force on the supporting column and, in return, the column presents an upward reaction to the beam it supports.

Calculation of reactions

The three rules of equilibrium can be used to calculate reactions. As discussed in Chapter 2 and again in Chapter 6, a reaction is a force (usually upwards) that occurs at a support of a beam or similar structural element. A reaction counteracts the (usually downward) forces in the structure to maintain equilibrium. It is important

Basic Structures, Second Edition. Philip Garrison.
© 2011 John Wiley & Sons, Ltd. Published 2011 by John Wiley & Sons, Ltd.

Total force up =

Total force down

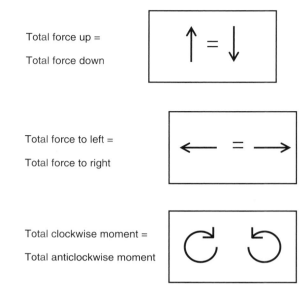

Total force to left =

Total force to right

Total clockwise moment =

Total anticlockwise moment

Fig. 9.1   The rules of equilibrium.

Fig. 9.2   Steel framed building under construction.
Note the cellular (i.e. with holes in them) steel beams, the steel columns and the profiled steel deck flooring.

to be able to calculate these reactions. If the support is a column, for example, the reaction represents the force in the column, which we would need to know in order to design the column.

Consider the example shown in Fig. 9.3. The thick horizontal line represents a beam of span 6 metres which is simply supported at its two ends, A and B. The only load on the beam is a point load of 18 kN, which acts vertically downwards at a

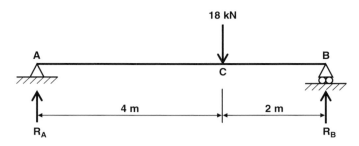

Fig. 9.3   Calculation of reactions for point loads.

position 4 metres from point A. We are going to calculate the reactions R_A and R_B (that is, the support reactions at points A and B respectively).

From *vertical equilibrium*, which we discussed in Chapter 6, we know that:

total force upwards = total force downwards

Applying this to the example shown in Fig. 9.3, we can see that:

$R_A + R_B = 18 \text{kN}$

Of course, this doesn't tell us the value of R_A and it doesn't tell us the value of R_B. It merely tells us that the sum of R_A and R_B is 18 kN. To evaluate R_A and R_B, we clearly have to do something different.

Let's use our new-found knowledge of moment equilibrium. We found out above that if any structure is stationary, then at any given point in the structure:

total clockwise moment = total anticlockwise moment

The above applies at any point in a structure. So, taking moments about point A:

$(18 \text{kN} \times 4 \text{m}) = (R_B \times 6 \text{m})$

Therefore $R_B = 12 \text{kN}$. Note that there is no moment due to force R_A. This is because force R_A passes straight through the point (A) about which we are taking moments.

Similarly, taking moments about point B:

total clockwise moment = total anticlockwise moment
$(R_A \times 6 \text{m}) = (18 \text{kN} \times 2 \text{m})$

Therefore $R_A = 6 \text{kN}$.

As a check, let's add R_A and R_B together:

$R_A + R_B = 6 + 12 = 18 \text{kN}$

which is what we would expect from the first equation above.

Fig. 9.4   Man on a scaffold board.

A word of warning

It's easy to make a mistake and get the two reactions the wrong way round. As a check, consider a man standing on a scaffold board supported by scaffold poles at each end, as shown in Fig. 9.4. The man is standing closer to the left-hand support. Which of the two supports is doing the more work in supporting the man's weight?

Common sense tells us that the left-hand support must be working harder to bear the man's weight, simply because the man is closer to that support. In other words, we would expect the left-hand support reaction to be the greater of the two.

Looking again at the example shown in Fig. 9.3, the 18 kN loading occurs towards the right-hand end of the beam, so we would expect the right-hand end reaction (R_B) to be greater than the left-hand end reaction (R_A). And indeed it is.

It's always a good idea to do this 'common sense check' to ensure that you've got the reactions the right way round. To summarise: if the loading on the beam is clearly greater at one end of the beam, you would expect the reaction to be greater at that end too.

Calculation of reactions when uniformly distributed loads are present

Up till now in this chapter we have looked only at problems with point loads and have studiously avoided those with uniformly distributed loads (UDLs). There is a good reason for this: analysis of problems with point loads only is much easier and it has been my policy in writing this book – as in life in general – to start with the easy things and work up to the harder ones.

In practice, most loads in 'real' buildings and other structures are uniformly distributed loads – or can be represented as such – so we need to know how to calculate end reactions for such cases. The main problem we encounter is in taking moments. For point loads it is straightforward – the appropriate moment is calculated by multiplying the load (in kN) by the distance from it to the point about which we're taking moments. However, with a uniformly distributed load, how do we establish the appropriate distance?

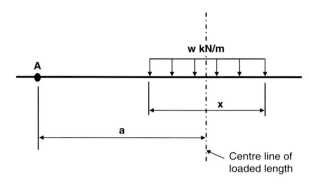

Fig. 9.5 Bending moment calculation for uniformly distributed load (UDL) general case.

Figure 9.5 represents a portion of uniformly distributed load of length x. The intensity of the uniformly distributed load is w kN/m. The chain-dotted line in Fig. 9.5 represents the centre line of the uniformly distributed load. Let's suppose we want to calculate the moment of this piece of UDL about a point A, which is located a distance a from the centre line of the UDL. In this situation, the moment of the UDL about A is the total load multiplied by the distance from the centre line of the UDL to the point about which we're taking moments. The total UDL is $w \times x$, the distance concerned is a, so:

moment of UDL about A $= w \times a \times x$.

Apply this principle whenever you're working with uniformly distributed loads.

Example involving uniformly distributed loads

Calculate the end reactions for the beam shown in Fig. 9.6. Use the same procedure as before.

Vertical equilibrium:

$$R_A + R_B = (3\,\text{kN/m} \times 2\,\text{m}) = 6\,\text{kN}$$

Taking moments about A:

$$(3\,\text{kN/m} \times 2\,\text{m}) \times 1\,\text{m} = R_B \times 4\,\text{m}$$

Therefore:

$$R_B + 1.5\,\text{kN}$$

Taking moments about B:

$$(3\,\text{kN/m} \times 2\,\text{m}) \times 3\,\text{m} = R_A \times 4\,\text{m}$$

Therefore:

$$R_A = 4.5\,\text{kN}$$

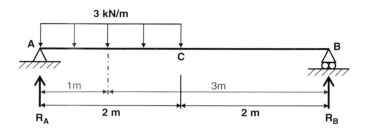

Fig. 9.6   Calculation of reactions for uniformly distributed loads.

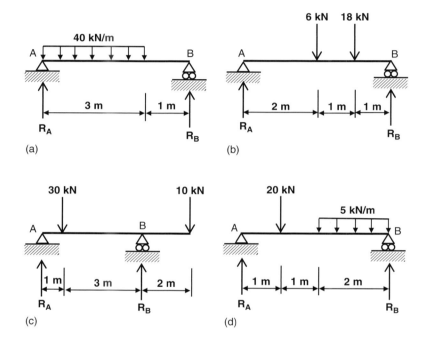

Fig. 9.7   Reactions – further tutorial examples.

Check:

$$R_A + R_B = 4.5 + 1.5 = 6\,\text{kN}$$

(as expected from the first equation).

What you should remember from this chapter

The third rule of equilibrium tells us that if an object, or any part of it, is stationary, the total clockwise moment about any point within the object is the same as the total anticlockwise moment about that point. This rule can be used in conjunction with the first two rules of equilibrium to calculate support reactions.

Tutorial examples

Try the examples given in Fig. 9.7. In each case, calculate the reactions at the support positions.

Tutorial answers

a) $R_A = 75\,\text{kN}$, $R_B = 45\,\text{kN}$
b) $R_A = 7.5\,\text{kN}$, $R_B = 16.5\,\text{kN}$
c) $R_A = 17.5\,\text{kN}$, $R_B = 22.5\,\text{kN}$
d) $R_A = 17.5\,\text{kN}$, $R_B = 12.5\,\text{kN}$

10 Different types of support – and what's a pin?

What is a pin?

I would be flattered to think that you are reading this book on a beach, at a rural beauty spot or in some other glamorous location. But the chances are you are inside a building as you skim through these words – perhaps your home, office or college – in which case, you will probably be in sight of a door. If it's a conventional door (not a sliding one, for example), it will have hinges on it. What are the hinges for? Well, they make it possible for you to open the door by rotating it about the vertical axis on which the hinges are located.

Figures 10.1a and 10.1b show the plan view of a door, in its shut and partially open positions respectively, along with part of the adjoining wall. You could approach this door and open it or shut it, partially or totally, at will. The hinges make it possible for you to do this by facilitating rotation. Had the door been rigidly fixed to the wall you would not have been able to open it at all. One other point to note: although you can open or shut the door at will, nothing you do to the door will affect the portion of the wall on the other side of the hinges. It remains unmoved. To put it another way, the hinges do not transmit rotational movements into the wall. This is a particularly important concept and is the basis of the analysis of pin-jointed frames, which we will investigate in Chapters 12–15.

The word *pin*, as used in structural engineering, is analogous to the hinge in a door. A pin is indicated symbolically as a small unfilled circle. Consider two steel rods connected by a pin joint, as shown in Fig. 10.2. The two rods are initially in line as shown in Fig. 10.2a and the left-hand rod is subsequently rotated about 30 degrees anticlockwise, as shown in Fig. 10.2b. The right-hand rod is not affected by this rotational movement of the left-hand rod.

A pin, then, has two important characteristics:

1) A pin permits *rotational* movement about itself.
2) A pin cannot *transmit* turning effects, or moments.

Different types of support

Up till now we've been talking about supports (to beams, etc.) and indicating them as upward arrows without giving any thought to the type or nature of the support. As we shall see, there are three different types of support: roller, pinned and fixed.

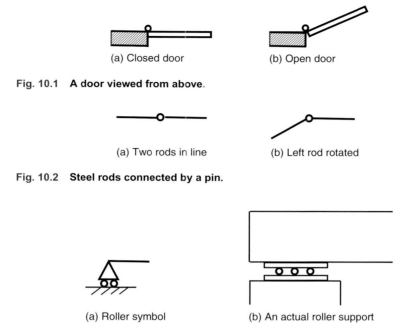

(a) Closed door   (b) Open door

Fig. 10.1   A door viewed from above.

(a) Two rods in line   (b) Left rod rotated

Fig. 10.2   Steel rods connected by a pin.

(a) Roller symbol   (b) An actual roller support

Fig. 10.3   Roller support – symbolically and in reality.

Roller supports

Imagine a person on roller skates standing in the middle of a highly polished floor. If you were to approach this person and give him (or her) a sharp push from behind (not to be recommended without discussing it with them first), they would move off in the direction you pushed them. Because they are on roller skates on a smooth floor, there would be minimal friction to resist the person's slide across the floor.

A ***roller support*** to part of a structure is analogous to that person on roller skates: a roller support is free to move horizontally. Roller supports are indicated using the symbol shown in Fig. 10.3a. You should recognise that this is purely symbolic and a real roller support will probably not resemble this symbol. In practice a roller support might comprise sliding rubber bearings, for example, or steel rollers sandwiched between steel plates, as shown in Fig. 10.3b.

Pinned supports

Consider the door hinge analogy discussed above. A ***pinned support*** permits rotation but cannot move horizontally or vertically – in exactly the same way as a door hinge provides rotation but cannot itself move away from its position in any direction.

Fixed supports

Form your two hands into fists, place them about a foot apart horizontally and allow a friend to position a ruler on your two fists so that it is spanning between them. Your fists are safely supporting the ruler at each end. Now remove one of the

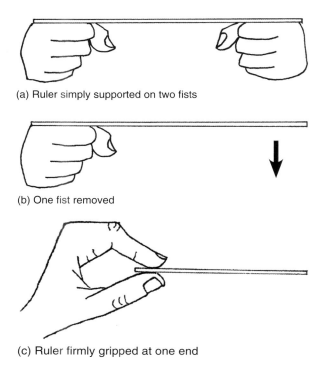

(a) Ruler simply supported on two fists

(b) One fist removed

(c) Ruler firmly gripped at one end

Fig. 10.4 What is a fixed support?

supports by moving your fist out from underneath the ruler. What happens? The ruler drops to the floor. Why? You have removed one of the supports and the remaining single support is not capable of supporting the ruler on its own – see Fig. 10.4a and b.

However, if you grip the ruler between your thumb and remaining fingers at one end only, it can be held horizontally without collapsing. This is because the firm grip provided by your hand prevents the end of the ruler from rotating and thus falling to the floor – see Fig. 10.4c.

In structures, the support equivalent to your gripping hand in the above example is called a *fixed support*. As with your hand gripping the ruler, a fixed support does not permit rotation.

There are many situations in practice where it is necessary (or at least desirable) for a beam or slab to be supported at one end only – for example, a balcony. In these situations, the single end support must be a fixed support because, as we've seen, a fixed support does not permit rotation and hence does not lead to collapse of the structural member concerned – see Fig. 10.5. Like a pinned support, a fixed support cannot move in any direction from its position. Unlike a pinned support, a fixed support cannot rotate. So a fixed support is fixed in every respect.

Now you've got a mental picture of each of the three different types of support (roller, pinned and fixed), let's revisit each of them and take our study of them a stage further. We are going to do this in the context of reactions and moments.

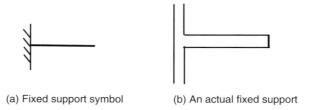

(a) Fixed support symbol   (b) An actual fixed support

Fig. 10.5   **Fixed support – symbolically and in reality.**

Restraints

Let's consider each of the following as being a restraint:

1) vertical reaction
2) horizontal reaction
3) resisting moment

Restraints experienced by different types of support

Roller support

Let's return to our roller skater standing on a highly polished floor. As the floor is supporting him, it must be providing an upward reaction to counteract the weight of the skater's body. However, we've already seen that if we push our skater, he will move. The rollers on the skates, and the frictionless nature of the floor, mean that the skater can offer no resistance to our push. In other words, the skater can provide no horizontal reaction to our pushing (in contrast to a solid wall, for example, which would not move if leaned on and therefore would provide a horizontal reaction). There is also nothing to stop the skater from falling over (i.e. rotating).

We can see from the above that a roller support provides one restraint only: ***vertical reaction.*** (There is no horizontal reaction and no moment.)

Pinned support

As discussed above, a pinned support permits rotation (so there is no resistance to moment), but because it cannot move horizontally or vertically there must be both horizontal and vertical reactions present. So, a pinned support provides two restraints: ***vertical reaction*** and ***horizontal reaction.*** (There is no moment.)

Fixed support

We saw above that a fixed support is fixed in every respect: it cannot move either horizontally or vertically and it cannot rotate. This means there will be both horizontal and vertical reactions and, if it cannot rotate, there must be a moment associated with the fixed support. Incidentally, this moment is called a ***fixed end moment*** – see Chapter 8 if you are not clear about the concept of a moment.

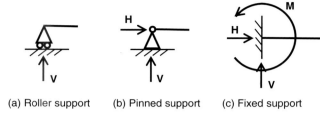

(a) Roller support   (b) Pinned support   (c) Fixed support

Fig. 10.6   Restraints provided by various support types.

So, a fixed support provides three restraints: vertical reaction, horizontal reaction and moment.

To summarise:

- A roller support provides one restraint: vertical reaction.
- A pinned support provides two restraints: vertical reaction and horizontal reaction.
- A fixed support provides three restraints: vertical reaction, horizontal reaction and moment.

This is illustrated in Fig. 10.6.

Simultaneous equations

Let's revise our knowledge of mathematics for a few minutes – specifically, equations and simultaneous equations.

Answer the following question with a simple Yes or No: can you solve the following equation?

$$x + 6 = 14$$

Clearly, the answer is Yes. You can solve the above equation very easily ($x = 8$), but why? The reason you can solve it so easily is that there is only one unknown (x in this case).

Now consider whether you can solve the following two simultaneous equations:

$$2x + 6y = -22$$
$$3x - 4y = 19$$

Again, it is possible to solve these two equations (although you may need to brush up your maths in order to do so). The solution, incidentally, is $x = 1$, $y = -4$. Again, why is it possible to solve these equations? The reason this time is that, although we have two unknowns (x and y), we have two equations.

Now consider whether or not you could solve the following simultaneous equations:

$$4x + 2y - 3z = 78$$
$$2x - y + z = 34$$

If you haven't realised for yourself, I'll spare you the tedium of trying to work it out by telling you: no, you can't solve the problem in this case. The reason is that this time we have three unknowns (x, y and z), but only two equations.

We could carry on investigating in this way for some time and if we were to do so we would find out the following:

- If we have the same number of unknowns as we have equations, a mathematical problem can be solved.
- However, if we have more unknowns than equations, a mathematical problem cannot be solved.

Relating this to structural analysis, if we look back to the procedure we used for calculating reactions in Chapter 9, we'll see that we were solving three equations. These equations were represented by:

1) Vertical equilibrium (total force up = total force down)
2) Horizontal equilibrium (total force right = total force left)
3) Moment equilibrium (total clockwise moment = total anticlockwise moment).

As we have three equations, we can use them to solve a problem with up to three unknowns in it. In this context, an unknown is represented by a restraint, as defined earlier in this chapter. (Remember, a roller support has one restraint, a pinned support has two restraints and a fixed support has three restraints.) So a structural system with up to three restraints is solvable – such a system is said to be *statically determinate* (SD) – while a structural system with more than three restraints is not solvable (unless we use advanced structural techniques which are beyond the scope of this book) – such a system is said to be *statically indeterminate* (SI).

So if we inspect a simple structure, examine its support and then count up the number of restraints, we can determine whether the structure is statically determinate (up to three restraints in total) or statically indeterminate (more than three restraints).

Let's look at the three examples shown in Fig. 10.7.

Example 1
This beam has a pinned support (two restraints) at its left-hand end and a roller support (one restraint) at its right-hand end. So the total number of restraints is $(2 + 1) = 3$, therefore the problem is solvable and is statically determinate.

Example 2
This pin-jointed frame has a pinned support (two restraints) at each end. So the total number of restraints is $(2 + 2) = 4$. As 4 is greater than 3, the problem is not solvable and is statically indeterminate.

Example 3
This beam has a fixed support (three restraints) at its left-hand end and a roller support (one restraint) at its right-hand end. So the total number of restraints is $(3 + 1) = 4$, therefore, again, the problem is not solvable and is statically indeterminate.

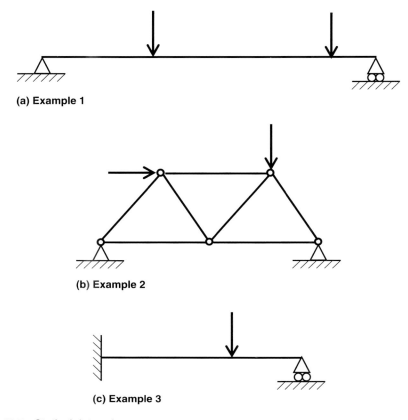

(a) Example 1

(b) Example 2

(c) Example 3

Fig. 10.7 **Statical determinacy.**

Fig. 10.8 **Lille Europe railway station, France.**

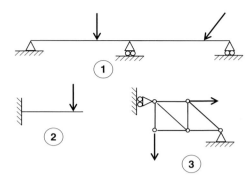

Fig. 10.9 **Statical determinacy – tutorial examples.**

What you should remember from this chapter

- Supports to structures are one of three types: roller, pinned or fixed. Each provides a certain degree of restraint to the structure at that point.
- Knowing the number of supports that a structure has and the nature of each support, it can be determined whether the structure is statically determinate (SD) or statically indeterminate (SI).
- A statically determinate structure is one that can be analysed using the principles of equilibrium discussed in the earlier chapters of this book. A statically indeterminate structure cannot be analysed using such principles.

Figure 10.8 shows Lille Europe railway station in Northern France. Look at the structure, most of which is helpfully painted light blue. Is it necessary to have arches in a building of this type, or are they there to make an otherwise ordinary structure look a little more interesting?

Tutorial examples

Determine whether each of the structures given in Fig. 10.9 is statically determinate (SD) or statically indeterminate (SI).

11 A few words about stability

Introduction

It is essential for a structure to be strong enough to be able to carry the loads and moments to which it will be subjected. But **strength** is not sufficient: the structure must also be **stable**.

In this chapter we'll be looking at what constitutes stability in structural terms – and how we can determine whether or not a particular structural framework is stable. Then we'll look, in practical terms, at how stability is achieved and ensured in buildings.

Stability of structural frameworks

Many buildings and other structures have a structural frame. Steel buildings comprise a framework, or skeleton, of steel. If you live in or near a large city you will have seen such frameworks being constructed. Many bridges also have a steel framework – famous examples include the Tyne Bridge in Newcastle upon Tyne and the Sydney Harbour Bridge in Australia.

We are going to consider the build-up of a framework from scratch. Our framework will consist of metal rods ('members') joined together at their ends by pins. (The concept of a pin, which is a type of connection that allows rotation, was discussed in Chapter 10.) Consider two members connected by a pin joint, as shown in Fig. 11.1a. Is this a stable structure? (In other words, is it possible for the two members to move relative to each other?) As the pin allows the two members to move relative to one another, this is clearly *not* a stable structure.

Now, let's add a third member connected by pin joints to form a triangle, as shown in Fig. 11.1b. Is this a stable structure? Yes, it is because even though the joints are pinned, movement of the three members relative to each other is not possible. So this is a stable, rigid structure. In fact, *the triangle is the most basic stable structure*, as we will mention again in the following discussion.

If we add a fourth member we produce the frame shown in Fig. 11.1c. Is this a stable structure? No it is not. Even though the triangle within it is stable, the 'spur' member is free to rotate relative to the triangle, so overall this is *not* a stable structure.

Basic Structures, Second Edition. Philip Garrison.
© 2011 John Wiley & Sons, Ltd. Published 2011 by John Wiley & Sons, Ltd.

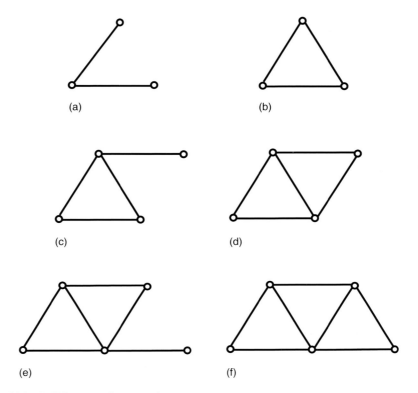

Fig. 11.1 Building up a framework.

Consider the frame shown in Fig. 11.1d, which is achieved by adding a fifth member. This is a stable structure. If you are unsure of this, try to determine which individual member(s) within the frame can move relative to the rest of the frame. You should see that none of them can, and therefore this is a stable structure. This is why you often see this detail in structural frames as 'diagonal bracing', which helps to ensure the overall stability of a structure.

Let's add yet another member to obtain the frame shown in Fig. 11.1e. Is this a stable structure? No, it is not. In a similar manner to the frame depicted in Fig. 11.1c, it has a spur member which is free to rotate relative to the rest of the structure. Adding a further member we can obtain the frame shown in Fig. 11.1f and we will see that this is a rigid, or stable, structure.

We could carry on in this vein, but I think you can see that a certain pattern is emerging. The most basic stable structure is a triangle (Fig. 11.1b). We can add two members to a triangle to obtain a 'new' triangle. All of the frames that comprise a series of triangles (Figs 11.1d and f) are stable; the remaining ones, which have spur members, are not.

Let's now see whether we can devise a means of predicting mathematically whether a given frame is stable or not. In Table 11.1 each of the six frames considered in Fig. 11.1 is assessed. The letter m represents the number of members in the frame and j represents the number of joints (note that unconnected free ends of members are

Table 11.1   Is a structure stable?

	m	j	Stable structure?	$2j - 3$	Is $m = 2j - 3$?
11.1a	2	3	No	3	No
11.1b	3	3	Yes	3	Yes
11.1c	4	4	No	5	No
11.1d	5	4	Yes	5	Yes
11.1e	6	5	No	7	No
11.1f	7	5	Yes	7	Yes

also considered as joints). The column headed 'Stable structure?' merely records whether the frame is stable ('Yes') or not ('No').

It can be shown that if $m = 2j - 3$ then the structure is stable. If that equation does not hold, then the structure is not stable. This is borne out by Table 11.1: compare the entries in the column headed 'Stable structure?' with those in the column headed 'Is $m = 2j - 3$?'.

Internal stability of framed structures – a summary

1) A framework which contains exactly the correct number of members required to keep it stable is termed a *perfect frame*. In these cases, $m = 2j - 3$, where m is the number of members in the frame and j is the number of joints (including free ends). Frames b, d and f in Fig. 11.1 are examples.
2) A framework having less than the required number of members is unstable and is termed a *mechanism*. In these cases, $m < 2j - 3$. Frames a, c and e in Fig. 11.1 are examples. In each case, one member of the frame is free to move relative to the others.
3) A framework having more than this required number is 'over-stable' and contains redundant members that could (in theory at least) be removed. Examples follow, and in these cases, $m > 2j - 3$. These frames are statically indeterminate (SI). We met this term in Chapter 10 – it means that the frames cannot be mathematically analysed without resorting to advanced structural techniques.

Examples
For each of the frames shown in Fig. 11.2, use the equation $m = 2j - 3$ to determine whether the frame is a) a perfect frame (SD), b) a mechanism (Mech) or c) statically indeterminate (SI). Where the frame is a mechanism, indicate the manner in which the frame could deform. Where the frame is statically indeterminate, consider which members could be removed without affecting the stability of the structure. The answers are given in Table 11.2.

The frames shown in Figs 11.2b, c and g are statically indeterminate. This means that they are over-stable and that one or more members could be removed without compromising stability. In the case of Fig. 11.2b, any one member could be removed from the top part of the frame and the structure would still be stable. In Fig. 11.2c, two

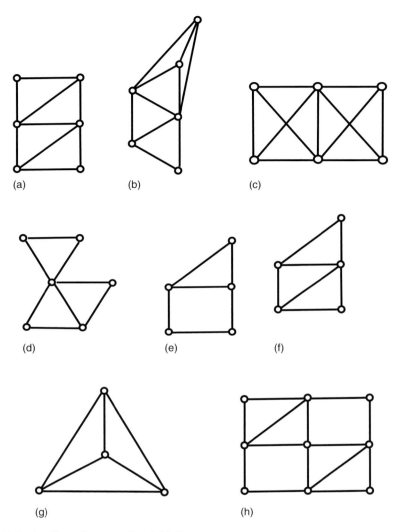

Fig. 11.2   Are these frameworks stable?

Table 11.2   Stability of frames shown in Fig. 11.2.

	m	*j*	*2j* – 3	Is *m* = 2*j* – 3? (or > or <)	Stability type
11.2a	9	6	9	=	SD
11.2b	10	6	9	>	SI
11.2c	11	6	9	>	SI
11.2d	8	6	9	<	Mech
11.2e	6	5	7	<	Mech
11.2f	7	5	7	=	SD
11.2g	6	4	5	>	SI
11.2h	14	9	15	<	Mech

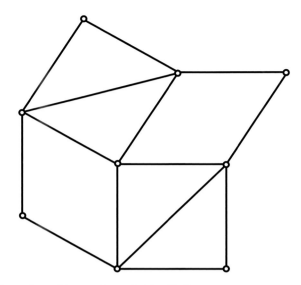

Fig. 11.3   Deformation of frame shown in Fig. 11.2h.

members could be removed without compromising stability – but the two members to be removed should be chosen with care. A sensible choice would be to remove one diagonal member from each of the two squares. In Fig. 11.2g, any one member could be removed.

The frames shown in Figs 11.2d, e and h are mechanisms. This means that a part of the frame is able to move relative to another part of the frame. In Fig. 11.2d, the upper triangle is free to rotate about the frame's central pin, independently of the lower part of the frame. In Fig. 11.2e, the square part of the frame is free to deform, or collapse, as we shall see in a later example.

The mode of deformation of the frame in Fig. 11.2h is less easy to visualise. It is shown in Fig. 11.3.

General cases

Look at the first two frames in Fig. 11.4. If we apply the $m = 2j - 3$ formula to the standard square depicted in Fig. 11.4a, we will find that it is unstable, or a mechanism. It can deform in the manner indicated by the broken lines. This is why, in 'real' structures, diagonal cross-bracing must often be provided to ensure stability.

If we look at the frame shown in Fig. 11.4b, we see that it is a square which is diagonally cross-braced twice. Applying the $m = 2j - 3$ formula we find that it is statically indeterminate, which means that it contains at least one redundant member. On further investigation we find that we can remove any one of the six members without affecting the stability of the structure.

During the years I've been teaching this subject I've discovered that many students derive a great deal of comfort from being taught a set of rules, or a 'magic formula', that could be applied to give the correct answer in any given situation.

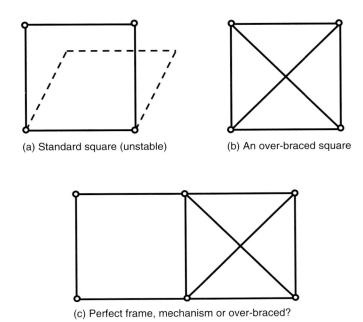

(a) Standard square (unstable)   (b) An over-braced square

(c) Perfect frame, mechanism or over-braced?

Fig. 11.4 Frame stability – general cases.

There is a tendency for them to regard such things as a crutch to be used as a substitute for analytical thought. Such students would readily latch on to the $m = 2j - 3$ formula discussed above as a universal panacea for determining the stability (or otherwise) of pin-jointed frames. I've got bad news for such readers: the above formula doesn't always work! (In fairness I should point out that there are other students who delight in finding the exception to the rule – and pointing it out to the lecturer.)

Consider the frame shown in Fig. 11.4c). It contains nine members and six joints, so $m = 9$ and $j = 6$ and it can thus readily be shown that $m = 2j - 3$ in this case, which suggests that the framework is a perfect frame. In fact, an inspection of the frame shows that this is not, in fact, the case. The left-hand part of the frame is an unbraced square, which is a mechanism and can deform in the same manner as the frame shown in Fig. 11.4a. But the right-hand part of the frame has double diagonal cross-bracing, which means that it is 'over-stable' and contains redundant members in the same way as the frame shown in Fig. 11.4b. So, part of the frame shown in Fig. 11.4c is a mechanism and the other part is statically indeterminate, but this does not make an overall perfect frame, as predicted by the formula!

The lesson to be learned from this is that the formula $m = 2j - 3$ should be regarded as a guide only – it doesn't always work. A given frame should always be inspected to see whether there are any signs of either a) mechanism or b) over-stability.

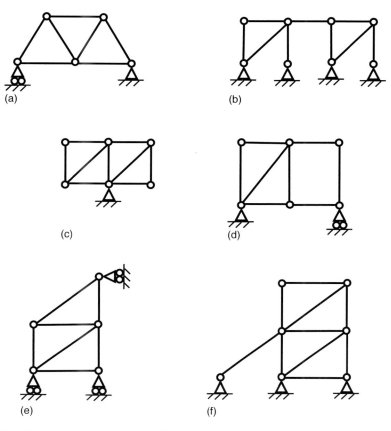

Fig. 11.5 **Are these structures stable?**

Frames on supports

So far in this chapter we have conveniently ignored the fact that, in practice, frames have to be supported. We therefore need to consider the effects of supports on the overall stability of frames.

In Chapter 10 we learned about the three different types of support (roller, pinned and fixed). We also saw that:

- a roller support provides one restraint ($r = 1$);
- a pinned support provides two restraints ($r = 2$);
- a fixed support provides three restraints ($r = 3$).

Reread Chapter 10 if you are unsure about this.

The $m = 2j - 3$ used above is now modified to $m + r = 2j$ where supports are present. As before, m is the number of members and j is the number of joints. The letter r represents the total number of restraints (one for each roller support, two for each pinned support and three for each fixed support).

Table 11.3   Stability of structures shown in Fig. 11.5

	m	*j*	*2j*	*r*	*m* + *r*	Is *m* + *r* = *2j*? (or > or <)	Stability type
11.5a	7	5	10	3	10	=	SD
11.5b	9	8	16	8	17	>	SI
11.5c	9	6	12	2	11	<	Mech
11.5d	8	6	12	3	11	<	Mech
11.5e	7	5	10	3	10	=	SD
11.5f	10	7	14	6	16	>	SI

- If $m + r = 2j$, then the frame is a perfect frame and is statically determinate (SD), which means that it can be analysed by the methods outlined in the following chapters of this book.
- If $m + r < 2j$, then the frame is a mechanism – it is unstable and should not be used as a structure.
- If $m + r > 2j$, then the frame contains redundant members and is statically indeterminate (SI), which means that it cannot be analysed without resorting to advanced methods of structural analysis.

Examples

For each of the frames shown in Fig. 11.5, use the equation $m + r = 2j$ to determine whether the frame is a) statically determinate, b) a mechanism or c) statically indeterminate. Where the frame is a mechanism, indicate the manner in which the frame could deform. Where the frame is statically determinate, consider which members could be removed without affecting the stability of the structure. The answers are given in Table 11.3.

The frames shown in Figs 11.5b and f) are statically indeterminate. This means they are over-stable and that one or more members may be removed. In the case of Fig. 11.5b, one of the diagonal members may be removed (but not both of them!) and the structure would still be stable. In Fig. 11.5f, the 'lean-to' diagonal member may be removed without compromising stability. The frames shown in Figs 11.5c and d) are mechanisms. The structure in Fig. 11.5c is obviously unstable, being free to rotate about its single central support. In Fig. 11.5d, the square part of the frame is free to deform in the manner indicated in Fig. 11.4a.

Stability of 'real' structures

In practice, the stability of a structure is assured in one of three ways:

1) shear walls/stiff core
2) cross-bracing
3) rigid joints

Let's look at each of these in more detail.

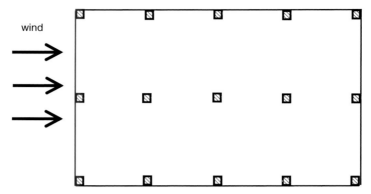

(a) Typical floor plan of reinforced concrete office building

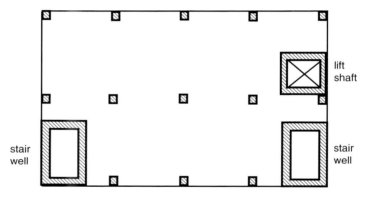

(b) Same floor plan with shear walls added

Fig. 11.6   Provision of stability using shear walls.

Shear walls/stiff core

This form of stability is usually (but not exclusively) used in concrete buildings. Consider the structural plan of an upper floor of a typical concrete office building, as shown in Fig. 11.6a. The structure comprises a grid layout of columns, which support beams and slabs at each floor level. The wind blows horizontally against the building from any direction. It is obviously important that the building doesn't collapse in the manner of a 'house of cards' under the effects of this horizontal wind force. We could design each individual column to resist the wind forces, but for various reasons this is not the way it is normally done.

Instead, *shear walls* are used. These walls are designed to be stiff and strong enough to resist all the lateral forces on the building. Since most buildings have staircases and many have lift shafts, the walls that surround the staircases and lift shafts are often designed and constructed to perform this role, as shown in Fig. 11.6b. On larger buildings, the shear walls may be constructed in such a way as to comprise

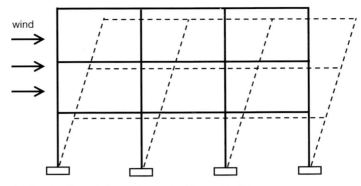

(a) Section through three-storey steel framed building

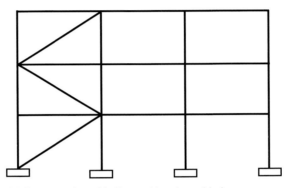

(b) Same section with diagonal bracing added

Fig. 11.7   Provision of stability using cross bracing.

an inner core to the building, which often contains stairwells, lift shafts, toilets and ducts for services. The NatWest Tower in London is an example of this form of construction.

Cross-bracing

This form of stability is common in steel-framed buildings. Figure 11.7a shows the elevation of a three-storey steel-framed building, on which the wind is blowing. There is nothing to stop the building tilting over and collapsing in the manner indicated by the broken lines.

One way of ensuring stability is to stop the 'squares' in the building elevation from becoming trapeziums. Earlier in this chapter we saw that a) a triangle is the most basic stable structure and b) a diagonal member can stop a square from deforming (illustrated in Figs 11.1b and d) respectively). So diagonal cross-bracing is used to ensure stability, as shown in Fig. 11.7b.

Large modern retail 'sheds', often occupied by do-it-yourself and electronics retailers, are found in most large British towns and cities. These are usually single-storey

Fig. 11.8   **Office building, Euston Road, London.**

steel structures and the structure of the building is often visible internally. Next time you visit such a store, have a look at its structure. You will notice steel columns at (typically) 5- or 6-metre intervals along the building. If you look at the end bay (i.e. the space between the end column and the next one) you may well see a zigzag arrangement of diagonal members. They are there for the reason discussed above: to provide lateral stability to the building as a whole. Figure 11.8 shows quite an 'extrovert' example of diagonal bracing on a new office building; note also the steel truss 'bridge' above the main entrance.

Rigid joints

A third method of providing lateral stability is simply to make the joints strong and stiff enough that movement of the beams relative to the columns is not possible. The black blobs in Fig. 11.9 indicate stiff joints that stop the action depicted in Fig. 11.7a from happening.

What you should remember from this chapter

- All structures must be stable, otherwise they may collapse. Being strong is not sufficient.
- A given structural framework may be either unstable, stable or over-stable. Which of these conditions applies can be determined through a combination of inspection and calculation.
- Lateral stability in buildings can be ensured through one of three means: shear walls, diagonal bracing or rigid connections.

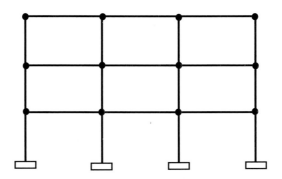

Fig. 11.9 **Provision of stability using rigid joints.**

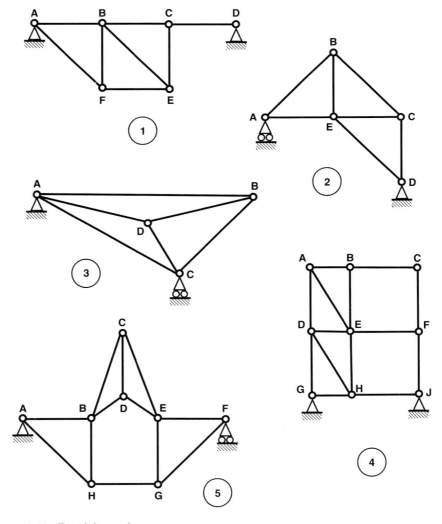

Fig. 11.10 **Tutorial questions.**

Tutorial examples

1) For each of the examples shown in Fig. 11.10, determine whether the frame is a) a perfect frame (SD), b) unstable (a mechanism) or c) over-stable (containing redundant members). If the framework is unstable, state where a member could be added to make it stable. If the frame is over-stable, determine which members could be removed and the structure would still be stable.

2) Select a framed structure near where you live. Determine how lateral stability is provided to the structure and state the reasons why the designer may have chosen that particular method of ensuring stability.

12 Introduction to the analysis of pin-jointed frames

Simple beams, lattice girders and trusses

The concept of a **beam** has been discussed in earlier chapters. We have seen that if a simple beam in a building is loaded from above, it will sag, as shown in Fig. 12.1. You can readily imagine that the material in the top of the beam is being squashed, or **compressed**. By contrast, the material in the bottom of the beam is being stretched – it is in **tension**.

The amount of downward movement, or **deflection**, from the horizontal depends in part on the material used – it is obviously a lot easier to bend a beam made of rubber than a beam of the same size made of timber.

Another factor that dictates the deflection of a beam is the shape and size of the beam's cross-section. If we consider a beam of rectangular cross-section, the shallower the beam is, the easier it is to bend. The reader can easily verify this point by gripping a plastic ruler at both ends and trying to bend it. If the ruler is orientated with its flat surface horizontal, it is easy to bend in a vertical plane. On the other hand, if the ruler is positioned 'on edge', it is very difficult to bend in a vertical plane, as shown in Fig. 12.2.

We can deduce from this that – all other things being equal – the deeper a beam is, the stronger it is. (This principle is demonstrated mathematically in Chapter 19.)

The problem is that, while a deep beam may be stronger than a shallow beam, it also requires more material, and material costs money. You might argue that use of more material is a price worth paying for a stronger beam, but there is a way round it. Instead of having a solid deep beam, it is possible to achieve the same result by having a framework of members, as shown in Fig. 12.3. The top and bottom members (or 'booms' as they are sometimes called) will be, respectively, in compression and tension, just as the top and bottom parts of a solid beam are. Such a framework is called a **lattice girder** or **truss** – it is usually made of steel but can be made of timber. You will have seen railway bridges that look like Fig. 12.3.

Other examples are shown in Figs 12.4–6. Figure 12.4 shows a modern lattice footbridge over a river; Fig. 12.5 illustrates a storey-depth lattice truss used in a building structure; and Fig. 12.6 shows the roof of a railway station where several different steel lattice girders are used for support.

Basic Structures, Second Edition. Philip Garrison.
© 2011 John Wiley & Sons, Ltd. Published 2011 by John Wiley & Sons, Ltd.

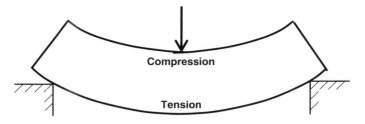

Compression

Tension

Fig. 12.1   **Bending of beams.**

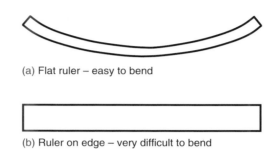

(a) Flat ruler – easy to bend

(b) Ruler on edge – very difficult to bend

Fig. 12.2   **Deeper beams are stronger.**

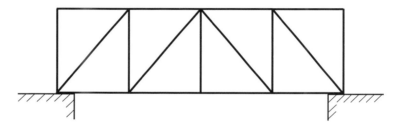

Fig. 12.3   **A steel railway bridge.**

Fig. 12.4   **Trussed bridge across Spree River, Berlin.**

Fig. 12.5   **Truss in facade, Sony Centre, Berlin.**

Fig. 12.6   **Roof structure, Manchester Victoria station.**

What is a pin-jointed frame?

Frameworks of structural members, such as steel railway bridges (as illustrated in Fig. 12.7) or pylons, are often analysed as pin-jointed frames. This means that the nodes, or joints, between members are regarded as pins or hinges, which by definition cannot transmit moments from one member to another. (See Chapter 10 for an explanation of the concept of a pin.)

It can be shown that the forces in the members of such frameworks are purely axial. In other words, the forces in the members act along the line of the members, which means that each member experiences one of the following:

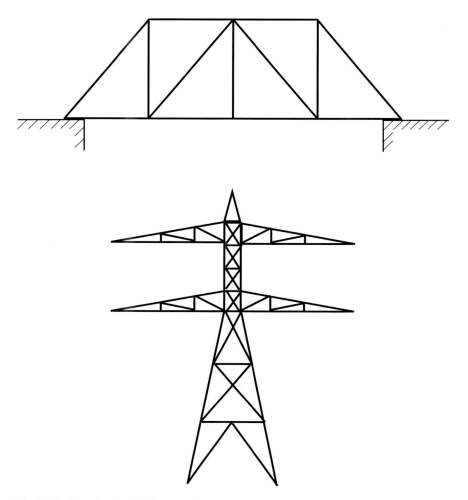

Fig. 12.7 Structural steel frameworks.

- pure compression
- pure tension
- no axial force

The members of a pin-jointed frame do not experience bending or shear forces.

You may question whether it is legitimate to analyse real structures as pin-jointed frames. After all, if you inspect the junction of two steel members in a railway bridge or an electricity pylon you will find that the junction is made of a combination of angle plates, bolts and welds, and may be quite complex – so surely it can't really be regarded as a pin joint?

Quite right – joints in structural frameworks are not usually pin-jointed in practice. But we consider the joints as pinned, for the purposes of analysis, for the following two reasons:

1) If you were to analyse the same structure a) assuming pin joints, then b) as a rigid-jointed structure, the results would be similar.
2) It is far easier to analyse the joints as being pins.

It follows therefore that we have to be able to analyse pin-jointed frames.

How are pin-jointed frames analysed?

By the term 'analysis', in the context of pin-jointed frames, we mean calculating:

1) the size of the *force* in each member
2) whether the force is *tensile* or *compressive*.

There are three techniques for doing so:

1) method of resolution at joints
2) method of sections
3) graphical method

These are discussed in Chapters 13, 14 and 15 respectively.

13   Method of resolution at joints

Introduction

The method of resolution at joints is the first of three alternative techniques for analysing pin-jointed frames. By 'analysing' we mean the process of calculating the force in each member of the pin-jointed frame and determining whether each of these forces is in tension or compression.

There are two other techniques:

- method of sections
- graphical (or force diagram) method

These two techniques are discussed in Chapters 14 and 15 respectively. The method of sections is appropriate only if the forces in some (rather than all) of the members are required. The graphical method, as its name suggests, involves scale drawing, which by its very nature introduces errors.

Students often have difficulty in understanding the techniques for analysis of pin-jointed frames. This is because these techniques are partly intuitive in nature. Because of these difficulties, students of architecture are often not taught how to analyse pin-jointed frames; if they receive any tuition on pin-jointed frames at all, it is usually merely conceptual. Some lecturers prefer to teach the graphical method to civil engineering students because a) it is non-mathematical and b) there is a rigid procedure to be followed, which makes it easier to teach and also easier for students to understand. However, the method of resolution at joints has more universal application and hence it will be taught in this chapter.

The rules

Throughout the analysis of pin-jointed frames using the method of resolution of joints, there are three rules to remember. These rules have all been taught in earlier chapters of this book and are as follows:

Rule 1: Force acts in same direction as member

The forces in any member of a pin-jointed frame are *axial*. In other words, the forces act along the centre line of a member. So, if a member is vertical, the forces in that member must be vertical. If a member is horizontal, the forces in that member will be

Basic Structures, Second Edition. Philip Garrison.
© 2011 John Wiley & Sons, Ltd. Published 2011 by John Wiley & Sons, Ltd.

horizontal. And if a member is inclined at an angle of, say, 30° to the horizontal, the forces within the member will act along that line.

Rule 2: Equilibrium applies everywhere

The basic rules of equilibrium apply at all nodes (and in all members) in a pin-jointed frame. This means that the sum of all downward forces on the node exactly equals the sum of all upward forces on the node. It also means that the total force to the left on the node exactly equals the total force to the right. See Chapter 6 if you are unclear on this point.

Rule 3: Forces can be split into components

If a force acts at an angle (i.e. it is neither horizontal nor vertical), then that force can be resolved into components – one horizontal and one vertical – which, taken together, have the same effect as the original force. Remember: if a force F acts at an angle θ to the horizontal, its horizontal component will always be $F.\cos \theta$ and its vertical component will always be $F.\sin \theta$ ('sign up'). See Chapter 7 if you need to review the concept of components.

Make sure that you fully understand the above three rules before proceeding, as they will come into play at every step in the following examples.

The general approach

As the term 'method of resolution at joints' implies, the technique involves examining each joint of a framework in turn. The easiest joints to analyse are those at which all forces and members are either horizontal or vertical. This is because, at such joints, there are no diagonal members – whose forces would have to be resolved into vertical and horizontal components.

Consider Fig. 13.1a, which shows the end part of a framework. No diagonal members radiate from corner B. The joint (or node) at this corner is subjected to a vertical force of 30 kN and a horizontal force of 64 kN as shown.

As the structure is presumably stationary, the rules of equilibrium will apply at the joint.

As total force up = total force down, then the vertical member of this framework (member AB) must experience a 30 kN upward force at point B (to oppose the external 30 kN downward force). Similarly, as total force to the left = total force to the right, then the horizontal member BD must experience a 64 kN rightward force at this point (to oppose the external 64 kN leftward force). See Fig. 13.1b.

Another thing to remember is that, just as joints must be in equilibrium, so too must members. In the horizontal member, we have a 64 kN force to the right; this must be opposed by a 64 kN force to the left at the other end of the member. In the vertical member there is a 30 kN force upwards; this must be opposed by a 30 kN force downwards at the other end of the member. See Fig. 13.1c. In the vertical member the arrows are pointing away from each other, so this member is in ***compression***. In the horizontal member the arrows are pointing towards each other, so this member is in **tension** (see Chapter 3 for reminder).

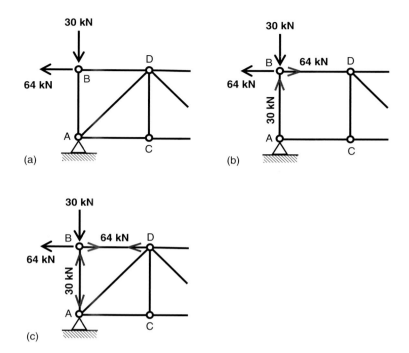

Fig. 13.1   Members in which forces are easily calculated.

Now look at the framework shown in Fig. 13.2a. It would take some time to analyse the whole frame, but there are certain members for which we could determine the forces straight away. Specifically, we could examine the joints at which there are no diagonal members or inclined forces, namely joints B, C and H.

Using the approach discussed above, we can see straight away that the force in member BD must be 12 kN (to oppose the horizontal 12 kN external force at B) and that it will be in compression (arrows pointing away from each other). Also, the force in member AB must be zero because there is no external vertical force to oppose at point B (or to put it another way, there is an external vertical force of 0 kN to oppose at point B).

Moving on to joint H, we see that the force in member GH must be 24 kN (to oppose the vertical 24 kN external force at H) and that it will be in tension (arrows pointing towards each other). The force in member FH must be zero because there is no external horizontal force at point H to be opposed (or to put it another way, there is an external horizontal force of 0 kN to oppose at point H).

Finally, let's look at joint C. The force in the vertical member CD must be zero because there is no external vertical force to oppose at point C. Furthermore, considering horizontal equilibrium at joint C, the forces in members AC and CE must be equal and opposite – although we cannot obtain their values without further analysis.

The forces we now know are shown in Fig. 13.2b.

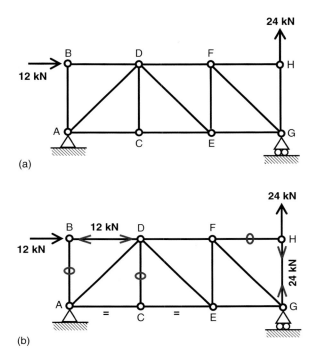

Fig. 13.2 **More members in which forces are easily calculated.**

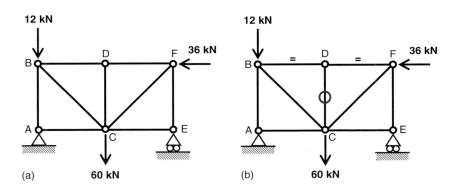

Fig. 13.3 **A case which is often misunderstood.**

Watch for the catch!

Now look at the frame shown in Fig. 13.3a. Looking at the frame, and without doing any calculation, what is the force in member CD?

I have presented this problem to students on numerous occasions. One common answer given to the above question is '60 kN'. I find this depressing because if you think the force in member CD is 60 kN, then I'm afraid you are wrong!

Look at joint D. There is no vertical external force there – or, if you prefer to think of it that way, the vertical external force at D is 0 kN. To balance this, the force in member CD must be 0 kN. The forces in the frame are shown in Fig. 13.3b. (Note that, for horizontal equilibrium at joint D, the forces in members BD and DF must be equal and opposite.)

So, why isn't the force in member CD 60 kN?

To answer this question we will look at joint C. Certainly, there is an external downward force of 60 kN there, which, for equilibrium, must be counteracted by an upward force of 60 kN. But member CD will not carry this vertical force alone: diagonal members BC and CF are also present at joint C and will carry a vertical component of force. Therefore the 60 kN upward force is shared between members BC, CD and CF – and, as we saw above, member CD actually carries no force in this case.

Standard cases

From the above discussion we can generate some standard cases of forces in certain members of pin-jointed frames. These standard cases are illustrated in Fig. 13.4.

In Fig. 13.4a, consideration of vertical equilibrium at joint A tells us that the force in member AB must be $F1$ to counter the vertical external force of $F1$ at joint A. (Note that the external force of $F3$ at joint B has no direct influence on the force in member AB.) Horizontal equilibrium at joint A tells us that the force in member AC – whatever it is – must be equal to the force in AD and, as the direction of the arrows must oppose each other at joint A for equilibrium, members AC and AD are either both in compression or both in tension. This is illustrated in Fig. 13.4b.

As there are no diagonal members present at joint E in Fig. 13.4c, the force in vertical member EG must be equal to the support reaction R. Furthermore, the force in horizontal member EH must be zero as there is no opposing horizontal external load. See Fig. 13.4d.

If we consider vertical and horizontal equilibrium at joint J in Fig. 13.4e we will see that the forces in members KJ and JL must both be zero as there are neither external forces nor diagonal members at joint J. This is shown in Fig. 13.4f.

You should realise by now that the force in member ST in Fig. 13.4g is not P. There are diagonal members present at joint S: the vertical components of the forces in these diagonal members will oppose the force P. The force in ST is in fact zero because there is no opposing external vertical force (or diagonal members to provide an opposing vertical force) at joint T. See Fig. 13.4h.

Study the standard cases shown in Fig. 13.4 and note particularly the presence or absence of diagonal members at the various joints.

The influence of diagonal members

It would make life easier, from the analysis point of view, if pin-jointed frames contained no diagonal members. Unfortunately, they always do: diagonal members are required to assure the frame's stability. So how do we analyse joints where diagonal members are present? Look at Fig. 13.5a, which shows a joint at the end of a frame. The joint comprises a horizontal member (AB) connected to a member

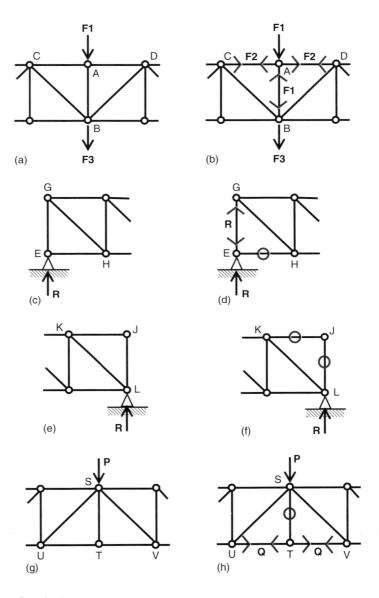

Fig. 13.4   Standard cases.

inclined at an angle of 60° to the horizontal (BC). A vertical external force of 3 kN acts at the joint. We wish to find the forces in members AB and BC.

If we resolve vertically at B, we can determine the force in member BC. The total force down at the joint (3 kN) will be equal to the total force up, which must be the vertical component of the force in member BC. So:

$$F_{BC}.\sin 60 = 3 \text{ kN, therefore } F_{BC} = 3.46 \text{ kN}$$

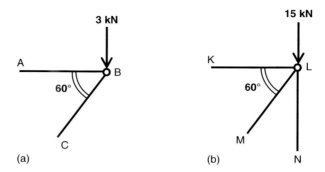

Fig. 13.5   Joints with diagonal members.

If we now resolve horizontally at B, we can calculate the force in member AB. The force in member AB will be equal to the horizontal component of the force in member BC. So:

$$F_{AB} = F_{BC}.\cos 60, \text{ therefore } F_{AB} = (3.46 \times 0.5) = 1.73\,\text{kN}$$

Now look at the joint L shown in Fig. 13.5b. We want to calculate the force in each of members KL, LM and LN, but it is not possible to do so from the information given: if we try to resolve either horizontally or vertically, we generate equations with more than one unknown, which cannot be solved. When analysing a frame with a joint like this, we should not start our analysis at this joint. Instead, we should start at another joint which resembles one of the examples above.

Now we will work through an entire framework in order to calculate all the forces in that frame. (Note: if the above calculation makes no sense at all to you, go back and read Chapter 7 – particularly the part on components.)

Worked example 1

See Fig. 13.6. The procedure is as follows:

1) Calculate the end reactions R_A and R_E in the same way as you would for a beam (see Chapter 9).
2) Proceed through the framework node by node, using the rules above to calculate the forces (and the directions of those forces) in each member.

Frequently asked question: How do I know which node to start at and which order to proceed through the nodes?

This is where the analysis becomes intuitive. You have to start at a node where there is not more than one unknown – but identifying such a node is not easy for the novice. Generally you should start at a support position, then move on to an adjacent node. The following example shows you how.

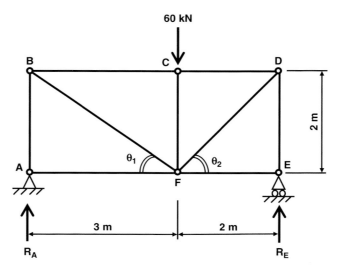

Fig. 13.6 **Worked example 1.**

Determination of reactions

From vertical equilibrium, the total force up ↑ = the total force down ↓. Therefore

$$R_A + R_E = 60\,\text{kN}$$

This doesn't tell us what R_A is; neither does it tell us what R_E is. It simply tells us that the two of them added together equals 60 kN. To evaluate R_A and R_E, we need another equation. This further equation can be determined from moment equilibrium – discussed in Chapter 6 – which tells us that the total clockwise moment about any stationary point is equal to the total anticlockwise moment about that point.

Considering moments about point A

Clockwise moment about point A due to external forces $= 60\,\text{kN} \times 3\,\text{m}$

Anticlockwise moment about point A due to external forces $= R_E \times 5\,\text{m}$

Equating these two:

$$R_E \times 5\,\text{m} = 60\,\text{kN} \times 3\,\text{m}$$

Therefore

$$R_E = 60\,\text{kN} \times 3\,\text{m}\,/5\,\text{m} = 36\,\text{kN}$$

Now since $R_A + R_E = 60\,\text{kN}$ (discussed above), then

$$R_A = 60 - 36 = 24\,\text{kN}$$

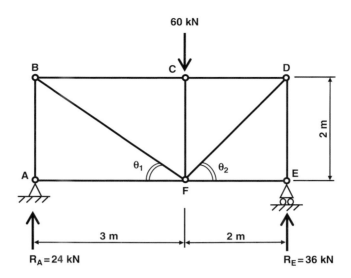

Fig. 13.7 Worked example 1 – with reactions calculated.

Applying the 'common sense check' (introduced in Chapter 9): the 60 kN load (which is the only load on the structure) acts to the right of centre, so it will be the right-hand support which 'does more work' in supporting the structure. Therefore we would expect the right-hand reaction (R_E) to be the greater of the two, which in fact it is (36 kN is greater than 24 kN).

Let's now add the reactions we've calculated to our diagram of the frame. See Fig. 13.7.

Analysis of the frame

Throughout this analysis, the following notation will be used:

F_{AB} represents the force in member AB
F_{BC} represents the force in member BC

… and so on.

Node A

There are three 'legs' to joint A:

- the vertical reaction R_A
- the vertical member AB
- the horizontal member AF

Resolving vertically at joint A

The term 'resolving vertically' means that we are considering the vertical forces (and vertical components of forces) associated with joint A, remembering that, for equilibrium, the total upward force at A has to be equal to the total downward force at A.

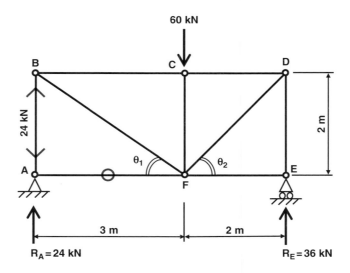

Fig. 13.8   **Worked example 1 – forces in members AB & AF calculated.**

Joint A experiences an upward force of 24 kN, in the form of the vertical reaction R_A. This means that, for equilibrium, there must be an (opposing) downward force of 24 kN at A. Since member AF, being horizontal, can contain only a purely horizontal force (i.e. no component of vertical force – see Rule 3 above), the downward 24 kN force can occur only in member AB. Therefore the force in member AB, F_{AB}, is 24 kN and is downwards in direction at end A.

Member AB

The principle of equilibrium applies in all parts of a structure or framework: not only at all nodes but in all members too. We have just determined that the force in member AB is 24 kN downwards at end A. As previously stated, wherever there is a downward force there must be an equal and opposite upward force, so it follows that there must be an upward force of 24 kN in member AB at end B.

Resolving horizontally at joint A

The term 'resolving horizontally' means that we are considering the horizontal forces (and horizontal components of forces) associated with joint A, remembering that, for equilibrium, the total horizontal force to the left at A is equal to the total horizontal force to the right at A.

The reaction at A, R_A, is purely vertical and has no horizontal component. Similarly, the force in member AB (which we now know to be 24 kN) is also purely vertical and has no horizontal component. Since there are no other external forces at joint A, the only member at joint A that can experience a horizontal force is member AF. And since there are no other horizontal forces to oppose it, the force in member AF, F_{AF}, must be zero.

Our framework now looks as shown in Fig. 13.8.

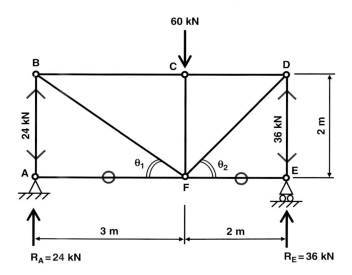

Fig. 13.9   Worked example 1 – forces in members DE & EF calculated.

We can now carry out a similar analysis of joint E. Using exactly the same approach as we used above for joint A, it can be shown that the force in member DE, F_{DE}, is 36 kN downwards (at end E) and the force in member FE, F_{FE}, is zero.

Our framework now looks as shown in Fig. 13.9.

Joint B

There are three 'legs' to node B:

- the vertical member AB (which contains a vertical force only)
- the horizontal member BC (which contains a horizontal force only)
- the inclined member BF (which, being inclined, will contain both horizontal and vertical components of force)

Resolving vertically at joint B

The only two members connecting at joint B that can have a vertical component of force are AB and BF. (Member BC, being horizontal, has no vertical force – see Rule 1 at the beginning of this chapter.)

We already know that there is an upward vertical force of 24 kN at joint B, contained in member AB. For equilibrium, there must be an opposing (downward) force of 24 kN and this must occur in member BF (i.e. the only other member at joint B that can contain a vertical force). Therefore the vertical component of the force in member BF must be 24 kN downwards.

Remembering that the vertical component of a force F at an angle θ is $F.\sin \theta$, it follows that in this case:

$$F_{BF} \times \sin \theta_1 = 24 \text{ kN}$$

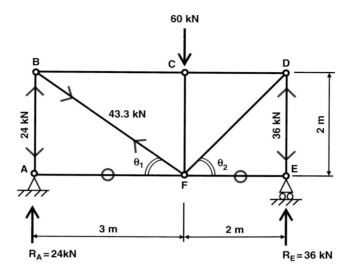

Fig. 13.10   Worked example 1 – force in member BF calculated.

Now θ_1 is the angle AFB $= \tan^{-1}(2/3) = 33.7°$. Therefore

$$F_{BF} \times \sin 33.7° = 24 \text{ kN}$$

So

$$F_{BF} = 24 / \sin 33.7° = 43.3 \text{ kN}$$

Let's now consider the direction of this force. We have said that the vertical component of the force in member BF (at end B) must act downwards. This means that the
force in member BF (at end B) must act downwards and to the right. Because equilibrium must apply in members as well as joints, this means that the force in member
BF at end F must oppose the force at end B; in other words, it must act upwards and
to the left.

Because the arrows in member BF point towards each other, member BF must be
in tension. (Remember from Chapter 3 that if the arrows in a member point towards
each other, that member is in *tension*. Think of the letter 'T' – the first letter of the
words 'towards' and 'tension'.)

The framework now looks as shown in Fig. 13.10.

Resolving horizontally at joint B

The only two members connecting at joint B that can have a horizontal component
of force are BF and BC. (Member BA, being vertical, has no horizontal force – again,
see Rule 1.) If the force in member BF (at end B) is 43.3 kN downwards and to the
right, then the horizontal component of this force is $F_{BF} \cos \theta_1 = 43.3 \times \cos 33.7° =$
36 kN (to the right).

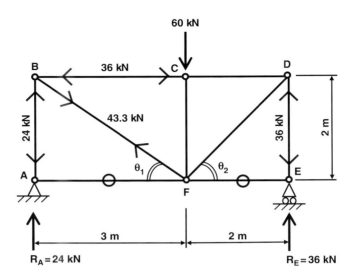

Fig. 13.11   Worked example 1 – forces in member BC calculated.

For equilibrium, there must be an opposing (to the left) force of 36 kN and this must occur in member BC (i.e. the only other member at joint B that can contain a horizontal force). So, the force in member BC (at end B) is 36 kN to the left. This will be opposed by a force of 36 kN to the right at end C. Therefore the two arrows in member BC point away from each other, so member BC must be in *compression*.

Our framework now looks as shown in Fig. 13.11.

We can now carry out a similar analysis of joint D. Using exactly the same approach as we used above for joint B, it can be shown that the force in member DF, F_{DF}, is 50.9 kN downwards and to the left (at end D) and the force in member DC, F_{DC}, is 36 kN to the right (at end D). (If you don't get those figures, remember that we have a different angle in this case: $\theta_2 = 45°$.)

Our framework now looks as shown in Fig. 13.12.

Joint C

Analysis of joint C is straightforward, as there are no inclined members to complicate matters.

Resolving vertically at joint C

There is a downward external vertical downwards force of 60 kN at joint C. To oppose this, the force in member CF (at end C) must be upwards. The force at the other end of CF will be downwards, therefore the member is in compression.

Resolving horizontally at joint C

The 36 kN force in member BC (at end C) is to the right, therefore, to oppose this, the force in member CD (at end C) must also be 36 kN, but to the left. The force at the other end of CD will be to the right, therefore the member is in compression.

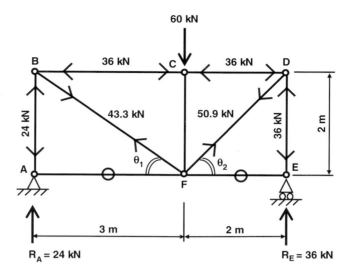

Fig. 13.12   Worked example 1 – forces in members DF & DC calculated.

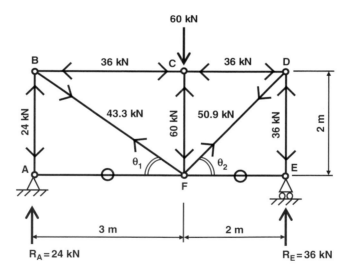

Fig. 13.13   Worked example 1 – frame fully analysed.

The framework now looks as shown in Fig. 13.13.

We have now established the magnitudes and directions of the forces in all the members. So have we finished this example? No, not quite. It would be prudent to carry out a check since, after all, it is quite possible that we may have made a mistake somewhere in our calculations. We can do this by resolving at a point not considered in our earlier calculations and checking, by calculation, that the forces previously calculated balance at that point.

Check: resolving vertically at joint F

As elsewhere, the total force up at joint F should equal the total force down. There are no external forces at joint F. The following members meet at joint F: AF, BF, CF, DF and EF. Members AF and EF are horizontal so can have no vertical forces (or vertical components of force) in them, so can be ignored when resolving vertically. This leaves members BF, CF and DF.

In our earlier calculations, we found that the vertical components of the forces in members BF and DF are upwards and we found that the vertical force in the (vertical) member CF is downwards. It follows, for equilibrium, that the sum of the vertical components of forces in members BF and DF (acting upwards) must equal the vertical force in member CF (acting downwards).

Vertical component of force in member:

$$BF = F_{BF} . \sin \theta_1 = 43.3 \times \sin 33.7° = 24 \text{ kN} \uparrow$$

Vertical component of force in member:

$$DF = F_{DF} . \sin \theta_2 = 50.9 \times \sin 45° = 36 \text{ kN} \uparrow$$

Vertical force in member:

$$CF = 60 \text{ kN} \downarrow$$

Since $24 + 36 = 60$, there is vertical equilibrium at joint F, so our earlier calculations are shown to be correct. A further check could be carried out by considering horizontal equilibrium at joint F.

Worked example 2

See Fig. 13.14. A different example, but the principles and the procedure are the same.

Determination of reactions

From vertical equilibrium, the total force up $\uparrow$ = the total force down $\downarrow$. Therefore

$$R_A + R_B = 200 \text{ kN}$$

Once again, this doesn't tell us what R_A is, neither does it tell us what R_B is. It simply tells us that the two of them added together equals 200 kN. To evaluate R_A and R_B, we need another equation. This further equation can be determined from moment equilibrium, discussed in Chapter 6, which tells us that the total clockwise moment about any stationary point is equal to the total anticlockwise moment about that point.

Considering moments about point A

Clockwise moment about point A due to external forces = 200 kN × 0.5 m

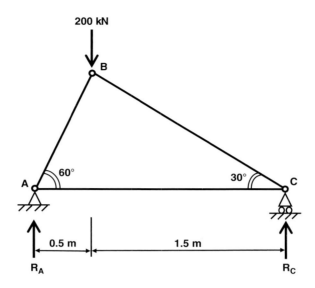

Fig. 13.14   Worked example 2.

Anticlockwise moment about point A due to external forces $= R_B \times 2\,\text{m}$
Equating these two:

$$R_B \times 2\,\text{m} = 200\,\text{kN} \times 0.5\,\text{m}$$

Therefore

$$R_E = 200\,\text{kN} \times 0.5\,\text{m} / 2\,\text{m} = 50\,\text{kN}$$

Now since $R_A + R_B = 200\,\text{kN}$ (discussed above), then

$$R_A = 200 - 50 = 150\,\text{kN}$$

Applying the 'common sense check': the 200 kN load (which is the only load on the structure) acts to the left of centre, so it will be the left-hand support which 'does more work' in supporting the structure. Therefore we would expect the left-hand reaction (R_A) to be the greater of the two, which in fact it is.

Let's now add the reactions we've calculated to our diagram of the frame. See Fig. 13.15.

Analysis of the frame

As before, the following notation will be used throughout this analysis:

F_{AB} represents the force in member AB
F_{BC} represents the force in member BC

... and so on.

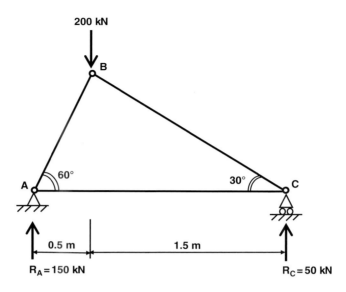

Fig. 13.15   Worked example 2 – with reactions calculated.

Node A

There are three 'legs' to joint A:

- the vertical reaction R_A
- the inclined member AB
- the horizontal member AC

Joint A experiences an upward force of 150 kN, in the form of the vertical reaction R_A. This means that, for equilibrium, there must be an (opposing) downward force of 150 kN at A. Since member AC, being horizontal, can contain only a purely horizontal force (i.e. no component of vertical force – see Rule 3 above), then the downward 150 kN force can occur only in member AB. Therefore the vertical component of the force in member AB is 150 kN. So

$$F_{AB} \times \sin 60° = 150 \text{ kN}$$

Therefore

$$F_{AB} = 150/\sin 60° = 173.2 \text{ kN}$$

which is downwards (and to the left) in direction at end A.

Member AB

As in the previous example, the downward (and to the left) force of 173.2 kN at end A of member AB must be opposed by an equal and opposite upward (and to the right) force of 173.2 kN at end B. (As the two arrows point away from each other, the member AB is in compression.)

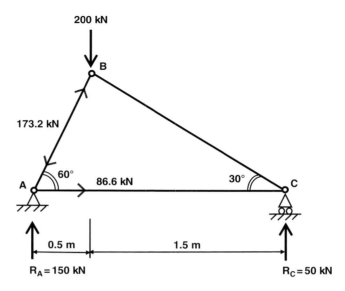

Fig. 13.16   Worked example 2 – forces in members AB & AC calculated.

Resolving horizontally at joint A

The term 'resolving horizontally' means that we are considering the horizontal forces (and horizontal components of forces) associated with joint A, remembering that, for equilibrium, the total horizontal force to the left at A is equal to the total horizontal force to the right at A.

The reaction at A, R_A, is purely vertical and has no horizontal component. But the force in member AB (which we now know to be 173.2 kN) is inclined and therefore will have a horizontal component. Member AC, being horizontal, will also experience a horizontal force. Since there are no other external forces at joint A, the force in member AC must be equal to the horizontal component of the force in member AB – but opposite in direction. So

$$F_{AC} = F_{AB} \times \cos 60°$$

But

$$F_{AB} = 173.2 \text{ kN (calculated above)}$$

Therefore

$$F_{AC} = 173.2 \times \cos 60° = 173.2 \times 0.5 = 86.6 \text{ kN}$$

Since the horizontal component of the force in member AB (at end A) acts to the left, the horizontal force in member AC (at end A) must act to the right.

Our framework now looks as shown in Fig. 13.16.

We can now carry out a similar analysis of joint C. Using exactly the same approach as we used above for joint A, it can be shown that the force in member CB, F_{CB}, is

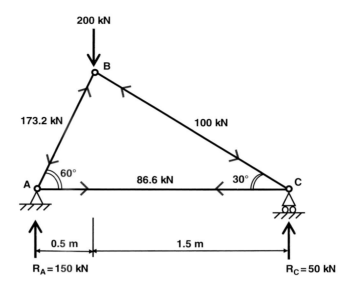

Fig. 13.17   Worked example 2 – frame fully analysed.

100 kN downwards and to the right (at end C) and the force in member BA, F_{BA}, is 86.6 kN to the left (at end C) which, as we would expect, exactly counteracts the force of 86.6 kN to the right at end A of that member. (Since the arrows in member AB point towards each other, the member is in *tension* – remember the letter 'T').

Our framework now looks as shown in Fig. 13.17.

We have now established the magnitudes and directions of the forces in all the members, but, as with the previous example, it would be wise to carry out a check by resolving at a point not considered in our earlier calculations and checking, by calculation, that the forces previously calculated balance at that point. If you were to resolve vertically at joint C, you should find that the forces at that joint balance.

Worked example 3

See Fig. 13.18 for a third and final worked example. After this you should be capable of working through the tutorial examples given at the end of the chapter. Remember that the same rules always apply. (Hint: you might find it helpful to do your own rough sketch of Fig. 13.18 and mark on the reactions and member forces as we calculate them.)

Determination of reactions

As always, we start by determining the reactions. If we consider the horizontal forces first, the only place we can have a horizontal reaction is joint A (the other support, F, is a roller and therefore cannot sustain a horizontal reaction). The horizontal reaction at A must be 20 kN (to the left) to oppose the only other horizontal force shown in Fig 13.18, which is a 20 kN horizontal force (to the right) at joint E.

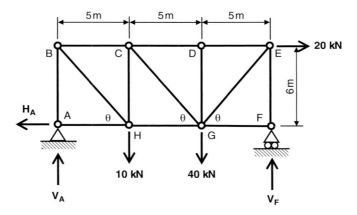

Fig. 13.18   Worked example 3.

Turning now to vertical reactions: each of the two supports will experience an upward vertical reaction: V_A at support A and V_F at support F. As always, total force up = total force down, so:

$$V_A + V_F = 10 + 40 = 50\,kN$$

Taking moments about support points

If we take moments about point A, we'll obtain an equation which we can solve for V_F. Don't forget to include the horizontal forces. The equation is:

$$(20\,kN \times 6\,m) + (40\,kN \times 10\,m) + (10\,kN \times 5\,m) = (V_F \times 15\,m)$$
Solving this gives $V_F = 38\,kN$.

If we take moments about point F, we'll obtain an equation which we can solve for V_A. I'll leave you to derive this equation for yourself – it will involve V_A and the 10, 20 and 40 kN forces, each of which should be multiplied by its distance from F – which can be solved to give $V_A = 12\,kN$.

'Easily calculated' forces

In line with the discussion earlier in this chapter, let's now identify those members in which the forces can be readily determined without calculation. In general, such members are attached to nodes from which there are no diagonal members. In the current example, such nodes are A, D and F, and we can easily deduce the forces in the following members:

- Member AH = 20 kN (tension) to oppose the 20 kN leftward horizontal reaction at A.
- Member AB = 12 kN (compression) to oppose the upward 12 kN vertical reaction at A.
- Member DG = 0 kN (no vertical forces at D). *Not 40 kN!*

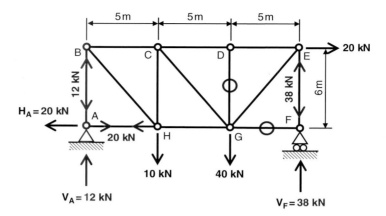

Fig. 13.19 Worked example 3 – with reactions calculated.

- Member CD and DE must have the same force in them (to maintain horizontal equilibrium at D), though we don't yet know the magnitude or sense (i.e. tensile or compressive) of that force.
- Member GF = 0 kN as there are no horizontal forces at F.
- Member EF = 38 kN (compression) to oppose the upward 38 kN vertical reaction at F.

The results of our work so far are shown in Fig. 13.19.

Node E

The forces in the remaining members can now be calculated using the method of resolution at joints. Remember, we normally start at the support nodes, but all the forces at A and F have already been determined. We could move on by analysing either node B or E; for no particular reason, I'm going to start at node E.

Let's start by calculating the angle θ shown on Fig. 13.18.

$\tan \theta = 6\,\text{m}/5\,\text{m} = 1.2$, so $\theta = 50.2°$
$\sin 50.2° = 0.768$ and $\cos 50.2° = 0.640$

Resolving vertically at E

$F_{GE} \times \sin 50.2° = 38\,\text{kN}$, so $F_{GE} = 38/0.768 = 49.5\,\text{kN}$ (tensile)

Resolving horizontally at E

Assuming the force in DE is compressive:

$F_{DE} + 20 = F_{GE} \times \cos 50.2°$, so $F_{DE} = 11.7\,\text{kN}$ (positive, so compressive as assumed)
Now let's move on to node G.

Resolving vertically at G

Assuming that the force in CG is tensile:

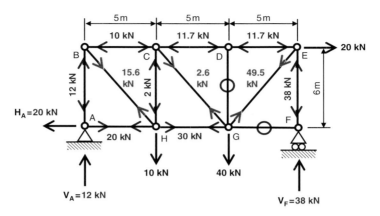

Fig. 13.20   Worked example 3 – frame fully analysed.

$F_{DG} + (F_{CG} \times \sin 50.2°) + (F_{GE} \times \sin 50.2°) = 40\,kN$
$0 + (F_{CG} \times 0.768) + (49.5 \times 0.768) = 40\,kN.$
So $F_{CG} = 2.6\,kN$ (positive, so tensile as assumed).

Resolving horizontally at G

Assuming the force in HG is tensile:

$F_{HG} + (F_{CG} \times \cos 50.2°) = (F_{GE} \times \cos 50.2°) + F_{GF}.$
$F_{HG} + (2.6 \times 0.640) = (49.5 \times 0.640) + 0$
So $F_{HG} = 30\,kN$ (positive, so tensile as assumed).
Now let's move on to node B.

Resolving vertically at B

$F_{BH} \times \sin 50.2° = 12\,kN$, so $F_{BH} = 15.6\,kN$ (tensile)

Resolving horizontally at B

$F_{BC} = F_{BH} \times \cos 50.2° = (15.6 \times 0.640) = 10\,kN$ (compressive)
  Now let's move on to node H.

Resolving vertically at H

Assuming the force in CH is compressive:

$(F_{BH} \times \sin 50.2°) - F_{CH} = 10\,kN.$
$(15.6 \times 0.768) - F_{CH} = 10\,kN.$
So $F_{CH} = 2\,kN.$

The only remaining force to calculate is in the member CD. But as we know that the forces in members CD and DE must be the same (horizontal equilibrium at D), then $F_{CD} = 11.7\,kN$ (compressive).
The fully analysed frame is shown in Fig 13.20.

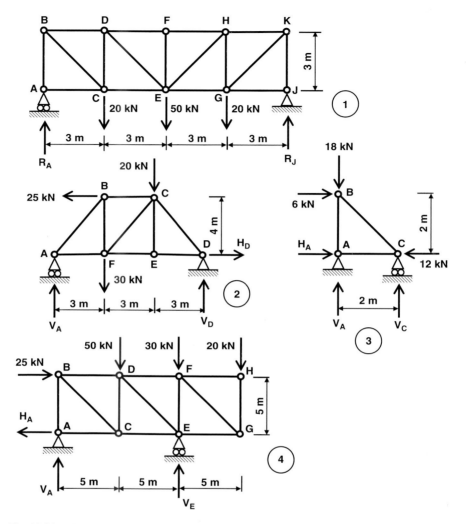

Fig. 13.21 **Method of resolution at joints – tutorial examples.**

Tutorial examples

Use the method of resolution at joints to find the forces in all the members of each frame given in Fig. 13.21.

14 Method of sections

Introduction

Sometimes we don't need (or wish) to determine the axial force in every member of a given pin-jointed frame, as we did when applying the method of resolution at joints in Chapter 13. We may wish to calculate the force in only one or two of the members. In such cases, the method of sections is useful.

In the method of resolution at joints, we doggedly worked our way through the structure, joint by joint, from one end of the structure to the other. As you will have found, this can get tedious, particularly when the structure has a large number of members and joints. In the method of sections, we establish a strategically placed 'cut line' through the structure. But determining the correct position of the cut line that will enable us to quickly solve the problem is crucial and partly intuitive, as we shall see.

To see why we might not want to work our way through a structure member by member, consider Figs 14.1 and 2.

Background to the method of sections

Imagine a steel framework that forms part of a railway bridge, as shown in Fig. 14.3a. Let's suppose that we wish to find the axial forces in members AB, BC and CD only. If the railway bridge was an existing structure and we were irresponsible enough to use suitable cutting tools to physically cut through the structure along a line through members AB, BC and CD, as shown in Fig. 14.3b, then what would happen? Obviously, the bridge would collapse.

Are there any circumstances under which the bridge would *not* collapse if cut through as shown? Well, collapse looks pretty inevitable, but there is one circumstance under which (in theory, at least) the bridge would not collapse.

- If it were possible to use some system of steel ropes, pulleys and props to provide exactly the same forces as existed in the members before they were cut, then the bridge would not collapse.

This means that if we could calculate the external forces in the cut structure that would keep that cut structure in overall equilibrium (indicated as F_{AB}, F_{BC} and F_{CD} in Fig. 14.3c), these would be the same as the internal forces that existed in members AB, BC and CD respectively before they were cut.

Basic Structures, Second Edition. Philip Garrison.
© 2011 John Wiley & Sons, Ltd. Published 2011 by John Wiley & Sons, Ltd.

Fig. 14.1   New York planetarium.
In this case of one structure (a spherical planetarium building) is encased inside another: a huge glass cube supported internally by steel lattice trusses.

Fig. 14.2   Swiss Re Building, London.
Popularly known as the 'gherkin' because of its distinctive shape, this was designed to provide maximum floor space with aerodynamic streamlining; architect Sir Norman Foster and engineer Ove Arup used an external 'diagrid' (steel members forming a series of triangles) to create the complex curved shape of the building.

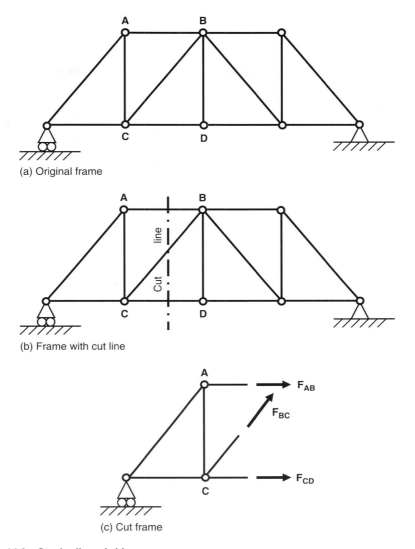

Fig. 14.3 Steel railway bridge.

To summarise then, the method of sections involves calculating the forces in certain members in a structure by pretending that the members concerned have been cut through and then calculating the external forces on the 'cut' structure. This process will be illustrated through the example that follows.

Example of method of sections

Suppose we wish to calculate the forces in members CD, HD and HG of the structure shown in Fig. 14.4a. We need to choose an appropriate cut line. In this case, a good choice would be a vertical cut line that passes through all three members, as shown in Fig. 14.4a.

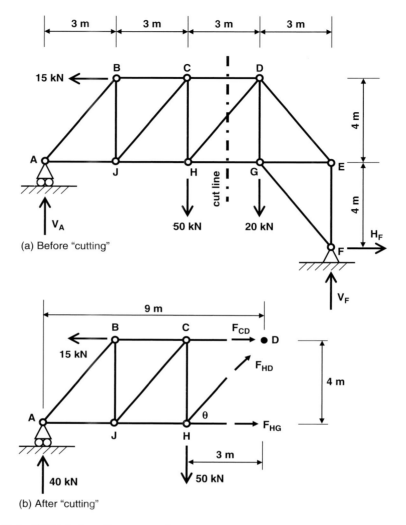

Fig. 14.4 **Method of sections example.**

First of all, we need to calculate the reactions in the usual way.

Calculation of reactions

From horizontal equilibrium of the whole structure:

$H_F = 15\,\text{kN}$ (i.e. Total force $\rightarrow$ = Total force $\leftarrow$)

From vertical equilibrium:

$V_A + V_F = 50 + 20 = 70\,\text{kN}$ (i.e. Total force $\uparrow$ = Total force $\downarrow$)

Taking moments about point A (i.e. Total clockwise moment = Total anticlockwise moment):

$$(50\,kN \times 6\,m) + (20\,kN \times 9\,m) = (V_F \times 12\,m) + (15\,kN \times 4\,m)$$
$$+ \ (15\,kN \times 4\,m)$$

So

$$V_F = 30\,kN$$

Taking moments about point F:

$$(V_A \times 12\,m) = (15\,kN \times 8\,m) + (50\,kN \times 6\,m) + (20\,kN \times 3\,m)$$

So

$$V_A = 40\,kN$$

If you don't follow the calculations above then I suggest you revisit the chapters on moments and reactions (Chapters 8 and 9).

The 'cut' section

Let us now suppose that we have cut the frame along the line shown in Fig. 14.4a. We will discard the part of the frame that is situated to the right of the cut line and will consider only the part to the left, as shown in Fig. 14.4b. If we can find the external forces F_{CD}, F_{HD} and F_{HG} that will keep this frame in equilibrium, these forces will correspond to the internal forces that existed in members CD, HD and HG (respectively) in the original pin-jointed frame.

Equilibrium of the frame shown in Fig. 14.4b

Considering vertical equilibrium:

$$40\,kN - 50\,kN + (F_{HD} \times \sin \theta) = 0 \text{ (i.e. Total force } \uparrow = \text{Total force } \downarrow)$$

You should realise that $(F_{HD} \times \sin \theta)$ is the vertical component of the force in member HD. Revisit Chapter 7 if you are unsure about this.
From basic trigonometry related to a right-angled triangle,

$$\tan \theta = 4\,m\,/\,3\,m = 1.333$$

Therefore $\theta = 53.1°$.
So if

$$40\,kN - 50\,kN + (F_{HD} \times \sin 53.1) = 0$$

this gives

$$F_{HD} = 12.5\,kN$$

We still need to find F_{CD} and F_{HG}. Let's take moments about point H. (Because the unknown force F_{HG} passes straight through point H, there will be no term involving F_{HG} in the equation if we use H as our 'pivot point' for taking moments. For the same reason, F_{HD} and the vertical 50 kN force at H will not come into the equation either.)

Taking moments about point H

(i.e. Total clockwise moment = Total anticlockwise moment)

$$(F_{CD} \times 4\,\text{m}) + (40\,\text{kN} \times 6\,\text{m}) = (15\,\text{kN} \times 4\,\text{m})$$

So

$$F_{CD} = -45\,\text{kN}$$

(The minus sign indicates that the force acts in the opposite direction to that assumed – so it acts to the left.)

The only remaining force to find is F_{HG}. Although we now know F_{CD} and F_{HD}, it would make life easier if we could take moments about the point through which both of these forces pass (i.e. point D) so there will be no term involving F_{CD} or F_{HD} (or, as it turns out, the 15 kN horizontal force at B). Note that it does not matter that point D is outside the frame we're considering: the rules of equilibrium hold for moments taken about any point, anywhere.

Taking moments about point D

(i.e. Total clockwise moment = Total anticlockwise moment)

$$(40\,\text{kN} \times 9\,\text{m}) = (50\,\text{kN} \times 3\,\text{m}) + (F_{HG} \times 4\,\text{m})$$

So

$$F_{HG} = 52.5\,\text{kN}$$

We've now calculated forces F_{CD}, F_{HD} and F_{HG}. We could check our calculations by considering horizontal equilibrium (i.e. Total force $\rightarrow$ = Total force $\leftarrow$) of the structure shown in Fig. 14.4b. But I'll leave that check to you …

So to summarise:

- The force in member CD is 45 kN and is compressive.
- The force in member HD is 12.5 kN and is tensile.
- The force in member HG is 52.5 kN and is tensile.

Summary of the method of sections

1) Calculate the end reactions in the usual way.
2) Decide in which member(s) you need to determine the force.

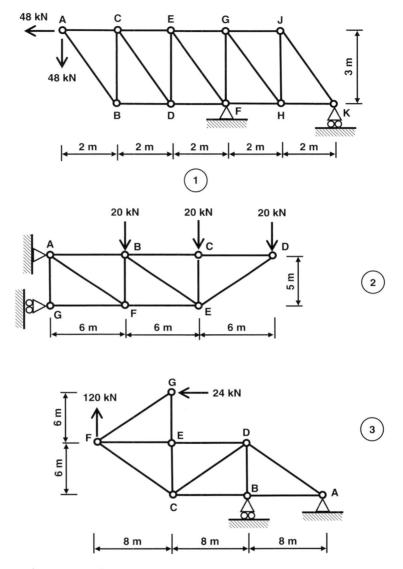

Fig. 14.5 Tutorial questions.

3) Draw a line that cuts through the member(s) of interest. (The cut line may be vertical, horizontal or inclined. It may be necessary to use different cut lines for different members.)
4) From now on, consider the part of the frame on one side of the cut line only (it doesn't matter which side).
5) Use the rules of equilibrium to determine the (now external) forces in the members of interest. Consider horizontal and/or vertical equilibrium and take moments about a strategically chosen point. These external forces correspond to the internal forces that existed in the members before they were 'cut'.

What you should remember from this chapter

This chapter outlines the method of sections. This is a useful procedure when we are interested in calculating the forces in only some (e.g. one or two) of the members of a pin-jointed frame. The concept involves pretending that the structure has been cut through the member concerned, then calculating the external forces that would be required to keep the 'cut' structure standing (i.e. in equilibrium). These external forces correspond to the external forces that existed in the 'cut' members before they were cut.

Tutorial examples

Use the method of sections to calculate the axial force and its sense (tension or compression) in the members stated below for each of the pin-jointed frames shown in Fig. 14.5:

- Frame 1: CD, DE, EG and GH.
- Frame 2: BE and BF.
- Frame 3: BC, CD and DE.

Check your answers using the method of resolution at joints (Chapter 13).

Tutorial answers

(All answers are in kN.)

- Frame 1: 57.6 (C), 48 (T), 144 (T), 129.8 (T)
- Frame 2: 62.5 (T), 60 (C)
- Frame 3: 296 (T), 200 (C), 112 (C)

15 Graphical method

Introduction

The previous two chapters discussed two methods of analysing pin-jointed frames, namely the method of resolution at joints and the method of sections. Both of these techniques are mathematical in nature, involving calculation. There is a third technique, called the graphical method (also known as the force diagram method). The graphical method is the subject of this chapter.

The graphical method involves no mathematical calculation whatsoever once the reactions have been calculated in the usual way. This, in itself, makes it appealing to some students. As the name suggests, the member forces and the type of forces are determined by constructing scale diagrams, for which you will need graph paper.

Example 15.1
The graphical method is best explained through example. The example we shall be working through in this chapter is illustrated in Fig. 15.1.

In the previous methods for pin-jointed frame analysis we labelled the joints. In the graphical method, we don't label the joints; instead, we label the areas or zones between the members of the frame and we do so in accordance with Bow's Notation, which is outlined below.

The graphical (force diagram) method in brief

1) Draw a load line for the applied loads and reactions, to scale. Start from the left-hand support.
2) Using Bow's Notation (see below), construct a force diagram, one joint at a time, drawing each line parallel to the direction of the member in the framework.
3) Load values can now be scaled from the diagram.
4) To determine the type of load in a member (i.e. tensile or compressive), 'travel' clockwise round a joint and note the force direction at the joint.
5) Construct a table.

Bow's Notation

Bow's Notation, named after its creator, is a convention for labelling the various zones in a diagram of a pin-jointed frame. Bow's Notation suggests the following:

Basic Structures, Second Edition. Philip Garrison.
© 2011 John Wiley & Sons, Ltd. Published 2011 by John Wiley & Sons, Ltd.

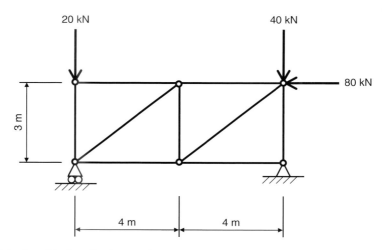

Fig. 15.1   Graphical method example.

1) Letter the spaces between the external applied loads and reactions.
2) Number the spaces between internal members.
3) Start with the letter 'A' between the reactions and work round the frame in a clockwise direction.
4) Start with the number 1 in the first left-hand space inside the framework.

If we label our frame in accordance with Bow's Notation, it will appear as shown in Fig. 15.2. Notice that the boundaries between the external zones (A, B, etc.) are defined by the positions of the lines of the external forces and reactions, and the members of the framework define the frontiers between the internal zones (1, 2, etc.).

As we progress through this problem, we will be constructing a diagram (called a *force diagram*) on a blank piece of graph paper. As we do this we will continually be referring back to the diagram shown in Fig. 15.2, which I will call the *frame diagram*.

Calculation of reactions

Let's start by calculating the reactions, which we will call V_L (vertical reaction, left-hand support), V_R (vertical reaction at right-hand support) and H_R (horizontal reaction, right-hand support).

From horizontal equilibrium,

$$H_R = 80\,\text{kN} \rightarrow$$

From vertical equilibrium,

$$V_L + V_R = 20 + 40 = 60\,\text{kN}$$

Taking moments about left-hand support:

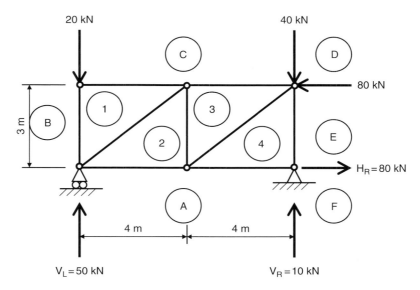

Fig. 15.2   Application of Bow's notation (frame diagram).

$$(40\,\text{kN} \times 8\,\text{m}) = (80\,\text{kN} \times 3\,\text{m}) + (V_R \times 8\,\text{m})$$

So

$$V_R = 10\,\text{kN}$$

and therefore

$$V_L = 50\,\text{kN}$$

Construction of the force diagram

We are now in a position to start constructing the force diagram. The various stages in the construction of this diagram are illustrated in Fig. 15.3.

Start with a blank piece of graph paper. Somewhere in the middle of the sheet, select a point and label it *a*. This (lowercase) *a* symbol on the force diagram corresponds to the (uppercase) zone A on the frame diagram. On the frame diagram (Fig. 15.2), you will notice that to get from zone A to zone B you need to cross a 50 kN upward force. This is represented on the force diagram by drawing a line vertically upwards from position *a* (representing zone A) for a distance representing 50 kN to arrive at a new position *b* (which represents zone B). To do this on graph paper, you will need to adopt a suitable scale – I would suggest that a scale of 1 mm = 1 kN would be suitable for this problem on an A4 sheet of graph paper.

So, the line 50 mm long, going up from point *a* to point *b* on the force diagram, represents the upward force (reaction) of 50 kN that you have to cross to get from zone A to zone B on the force diagram.

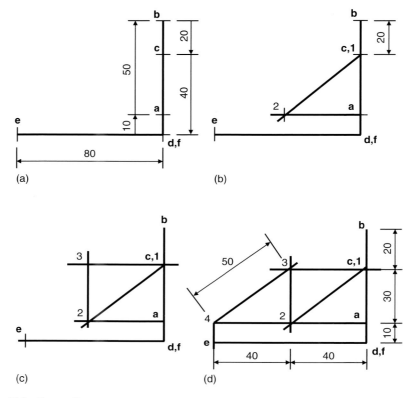

Fig. 15.3   Force diagram.

Returning to the frame diagram, getting from zone B to zone C entails crossing a 20 kN downward force (see Fig. 15.2). On the force diagram (Fig. 15.3) this is represented by drawing a line vertically downwards from position *b* of length 20 mm (equivalent to 20 kN). The point arrived at is labelled *c* and represents zone C on the frame diagram.

Back with the frame diagram again, it can be seen that:

- getting from zone C to zone D involves crossing a 40 kN force (vertically downwards)
- getting from zone D to zone E involves crossing an 80 kN force (to the left)
- getting from zone E to zone F involves crossing an 80 kN force (to the right)
- getting from zone F to zone A involves crossing a 10 kN force (vertically upwards).

These are represented, respectively, by:

- a vertical line downwards from *c*, 40 mm long, to establish *d*
- a horizontal line leftwards from *d*, 80 mm long, to establish *e*

- a horizontal line rightwards from *e*, 80 mm long, to establish *f*
- a vertical line upwards from *f*, 10 mm long, to establish *a*.

The resultant force diagram is shown in Fig. 15.3a.

The next task is to locate the points 1, 2, 3 and 4 on the force diagram, which respectively represent zones 1, 2, 3 and 4 on the frame diagram. Examine zone 1 on the frame diagram (Fig. 15.2). It is separated from zone B by a vertical member and from zone C by a horizontal member. This dictates that, on our force diagram:

- point 1 lies on a vertical line that also passes through point *b*
- point 1 lies on a horizontal line that also passes through point *c*.

So point 1 (representing zone 1) must lie at the point shown on Fig. 15.3b.

Moving on to zone 2 on the frame diagram, it can be seen that this is separated from zone A by a horizontal member and from zone 1 by a diagonal line sloping upwards and to the right at an angle of '4 squares along, 3 squares up' (or 36.9°). So point 2 can be found on our force diagram from the following two rules:

- point 2 lies on the diagonal line (angle as above) that also passes through point 1
- point 2 lies on a horizontal line that also passes through point *a*.

So point 2 must lie at the point shown on Fig. 15.3b.

By a similar process, point 3 lies at the point where a vertical line through point 2 intersects a horizontal line through point *c* (see Fig. 15.3c) and point 4 lies at the point where a vertical line through point *e* meets a horizontal line through point *a*. The completed force diagram is shown in Fig. 15.3d.

Using the force diagram to determine the magnitude of forces

Now comes the easy bit. To determine the force in a member, you simply scale off the distance between the relevant two points on the force diagram (Fig. 15.3d). For example, to determine the force in the right-hand diagonal member of the framework, which separates zone 3 from zone 4 on the frame diagram (Fig. 15.2), you need to measure the distance between points 3 and 4 on the force diagram (Fig. 15.3d). This distance is 50 mm and thus the force in the member is 50 kN. (Note: Please do not scale off the diagrams in this book as they are not to the correct scale, but your own force diagram will be.)

Similarly, to determine the force in the central vertical member, which separates zones 2 and 3 on the frame diagram, it is necessary to measure the distance between points 2 and 3 on the force diagram. It can readily be seen from Fig. 15.3d that this distance is 30 mm and thus the force in the member is 30 kN.

If you were to carry out this process for the remaining members, the forces you would obtain are shown in Table 15.1.

We're now half way to solving this problem. We have worked out the magnitudes of the forces in each member. Keep reading to find out how we determine the type of force (tension or compression) in each member.

Table 15.1   Member forces in Example 15.1

Member reference	Axial force in member (kN)
B–1	20
C–1	0
1–2	50
A–2	40
2–3	30
C–3	40
3–4	50
A–4	80
E–4	10

The member references represent the zones (as shown in Fig. 15.2) that the member lies between. For example, member B–1 lies between zones B and 1 (i.e. the left-hand vertical member), member 3–4 lies between zones 3 and 4 (i.e. the right-hand inclined member) and so on.

The van driver analogy

Imagine you are a delivery van driver, based in the town of Mitchellstown. On a particular day, you have to make deliveries to addresses in three different towns: Pennyport, Jackston and Charlesville. It is up to you to decide the order in which you visit the three towns. The highway system linking the three towns to each other and Mitchellstown is shown in Fig. 15.4a. From this you can see that the most efficient two options are:

1) Mitchellstown – Pennyport – Jackston – Charlesville – Mitchellstown (i.e. a clockwise circuit).
2) Mitchellstown – Charlesville – Jackston – Pennyport – Mitchellstown (i.e. an anticlockwise circuit).

You are trying to decide which of the options to go with when your boss comes running out of his office. He tells you he has had an urgent phone call and asks you to do the Charlesville delivery first. So the decision is made for you: you need to visit the towns in the order given in the second option above, shown in Fig. 15.4b.

Calculation of the sense (compressive or tensile) of the internal forces in the framework

Returning to the example presented in Fig. 15.1, we have now drawn our force diagram (Fig. 15.3), from which we have scaled off the magnitude of the forces (presented in Table 15.1). But how do we determine which of these forces are in tension and which are in compression?

Joint at top right-hand corner of frame

Consider the top right-hand corner of the frame in our example. By inspection of Fig. 15.2, it can be seen that five zones meet at this point. (If it helps, and if you've got

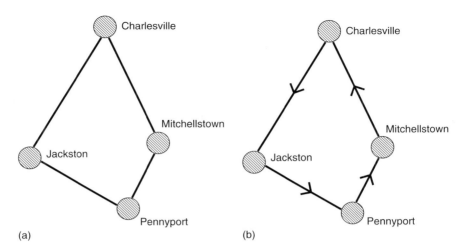

(a)   (b)

Fig. 15.4   A van driver's delivery route

agricultural interests, it might help to consider this point as the place where five fields meet and you have to name them.) The five zones meeting at this point are: C, D, E, 3 and 4.

If we now turned to the force diagram (Fig. 15.3d) and superimposed thick lines on it representing the links between these five points (c, d, e, 3 and 4) we would end up with the diagram shown in Fig. 15.5a. Now we know that a 40 kN downward force separates zones C and D (Fig. 15.2), so this can be represented by a downward arrow between points c and d in Fig. 15.5a. Similarly, the 80 kN leftward force separating zones D and E can be represented by a leftward arrow between points d and e in Fig. 15.5a.

By reference to the van driver's analogy above, the directions of these two forces determine the directions of the other forces to complete the circuit in Fig. 15.5a – shown by arrowheads. So the force in line e–4 is upwards, 4–3 is upwards and rightwards and in line 3–c is rightwards, as shown in Fig. 15.5a. If we transfer these force directions to the corresponding members of the frame diagram we see that the direction of the forces on the frame diagram will be as shown in Fig. 15.5b.

Joint at bottom left-hand corner of frame

Now let's consider the bottom left-hand corner of the frame. Looking at Fig. 15.2, it can be seen that four zones meet at this point, namely A, B, 1 and 2. If we now turned to the force diagram (Fig. 15.3d) and superimposed thick lines on it representing the links between these four points (a, b, 1 and 2) we would end up with the diagram shown in Fig. 15.5c.

Now we know that a 50 kN upward force separates zones A and B (see Fig. 15.2), so this can be represented by an upward arrow between points a and b in Fig. 15.5c.

Once again, the direction of this force determines the directions of the other forces to complete the circuit in Fig. 15.5c – shown by arrowheads. This tells us that the force in line b–1 is downwards, 1–2 is downwards and leftwards, and 2–a is rightwards. So the direction of the forces on the frame diagram will be as shown in Fig. 15.5d. Repeating the process for every joint will give the arrow formation shown in Fig. 15.6. Remember:

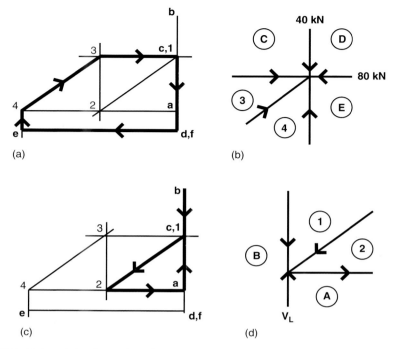

Fig. 15.5 **Determining force directions.**

- arrows pointing towards each other indicate *tension*
- arrows pointing away from each other indicate *compression*.

So, to sum up, the procedure for determining which members are in tension and which are in compression is as follows:

1) Consider each joint in turn.
2) For the chosen joint, consider which zone numbers/letters directly contact the joint.
3) Draw a thick line connecting the corresponding zone numbers on the force diagram.
4) The direction of the force between two of the zone numbers is usually known. From this the direction of all the other forces can be determined.

The roof shown in Fig. 15.7, which was photographed from the platform of an underground railway station many metres below, is a typical space frame. A space frame is a three-dimensional pin-jointed frame and has to be designed accordingly.

What you should remember from this chapter

This chapter describes the graphical method, which is a procedure for determining the forces in pin-jointed frames using drawing rather than calculation. The procedure can best be learned by following the example used in this chapter and applying it to the tutorial examples given in the following section.

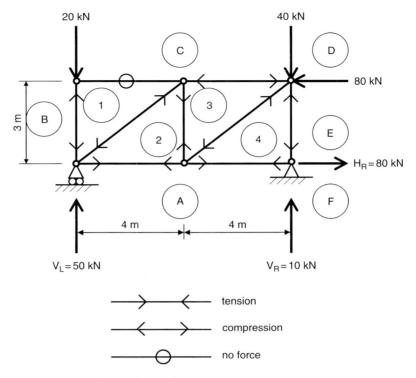

Fig. 15.6   Direction of forces in members.

Fig. 15.7   Steel space frame roof, Lille Europe metro station, France.

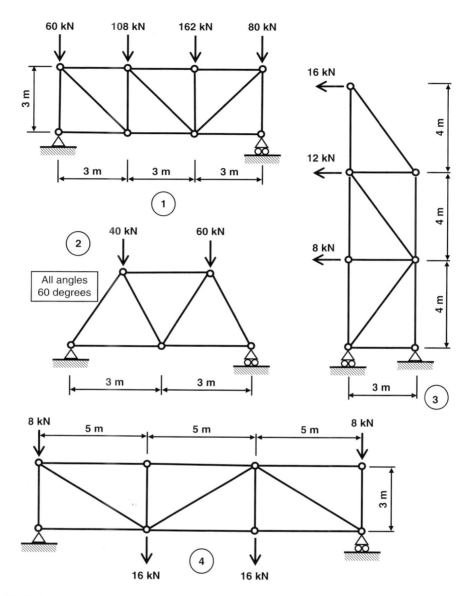

Fig. 15.8   Tutorial questions.

Tutorial examples

Use the graphical method to determine the forces in each member of each of the examples illustrated in Fig. 15.8. In each case, find out whether the force is tensile or compressive. Then either check your answers using the method of resolution at joints (Chapter 13) or check the forces in selected members using the method of sections (Chapter 14).

16   Shear force and bending moments

Introduction

We encountered the concepts of shear and bending in Chapter 3. In this chapter these concepts are explored further and their quantification and calculation are explained.

Deformation of structures

Imagine that the beams indicated by the thick solid horizontal lines in Fig. 16.1 are quite flexible but not particularly strong, so will readily deform under the loads shown. The lines in Fig. 16.2 indicate the deformed (or deflected) forms of the corresponding beams in Fig. 16.1.

Hogging and sagging

We're going to discuss the deformations shown in Fig. 16.2, but before we do, let's define two important terms. You have probably already encountered the term *sagging* – for example, you may have a bed that sags, or dips, in the middle (in which case, my advice is: get a better bed – it's well worth the investment). Sagging, or downward deformation, is illustrated in Fig. 16.3a.

Hogging – an upward deformation – is the opposite (or mirror image) of sagging. The concept of hogging is illustrated in Fig. 16.3b.

Discussion of the deflected forms shown in Fig. 16.2

Consider, as an example, beam 1 in Fig. 16.1, which is simply supported at either end and is subjected to a central point load. Clearly, the beam will tend to sag under that load, as indicated by the line in the corresponding diagram in Fig. 16.2. When the beam has sagged, the fibres in the very top of the beam will be squashed together; in other words, they will be compressed. Similarly, the fibres in the bottom part of the beam will have stretched, which indicates that the bottom of the beam is in tension. The fact that the bottom of the beam is in tension is indicated by the letter T (for tension) placed underneath the line in beam 1 in Fig. 16.2.

Beam 2 in Fig. 16.1 will tend to hog (or 'break its back') over the central support as a result of the point loads at either end. This hogging profile is indicated by the line

Basic Structures, Second Edition. Philip Garrison.
© 2011 John Wiley & Sons, Ltd. Published 2011 by John Wiley & Sons, Ltd.

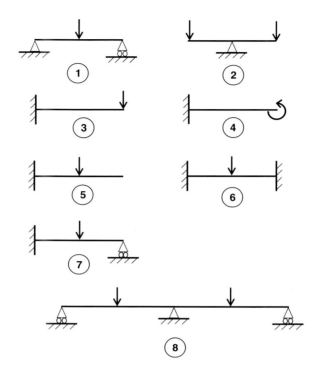

Fig. 16.1 Deformations in beams.

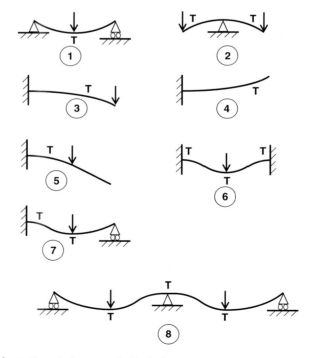

Fig. 16.2 Deformations in beams – indicated.

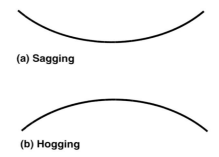

(a) Sagging

(b) Hogging

Fig. 16.3   **Hogging and sagging.**

in the corresponding diagram in Fig. 16.2. In this case, we will see that the top of the beam will be in tension and therefore we've indicated tension (letter T) above the line at the support position.

We can analyse the remaining beams in Fig. 16.1 in a similar fashion and obtain the deformed profiles and tension positions for each one (indicated by the lines and letter T respectively in Fig. 16.2).

If you have difficulty visualising the deformation of the beam shown in beam 4, replicate the situation by holding a standard-length ruler horizontal by gripping it firmly with your left hand at its left-hand end and applying an anticlockwise twist with your right hand at the right-hand end. You will then see the ruler deform in the manner depicted for beam 4 in Fig. 16.2 and tension will occur on the underside.

When examining the deformed shapes of the beams indicated in Fig. 16.2 for beams 6 and 7, remember that a fixed support firmly grips a beam, while a pinned (or simple) support permits rotation to take place. (See Chapter 10 to remind yourself of the various support types.)

If you completely understand Fig. 16.2, move on to the next section.

Shear and bending

You were introduced to the concepts of *shear* and *bending* in Chapter 3. These two terms represent the ways in which a structural member (for example, a beam) can fail, and this was illustrated in Figs 3.4 and 3.5. To remind you:

- Shear is a cutting or slicing action which causes a beam to simply break or snap. As discussed in Chapter 3, a heavy load located near the support of a weak beam might cause a shear failure to occur.
- If a beam is subjected to a load it will bend. The more load that is applied, the more the beam will bend. The more the beam bends, the greater will be the tensile and compressive stresses induced in the beam. Eventually, these stresses will increase beyond the stresses that the material can bear and failure will occur – in other words the beam will break. In short, if you increase the bending in a beam, eventually it will break.

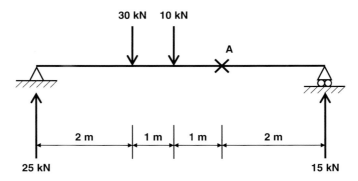

Fig. 16.4   Example 16.1 – Shear force and bending moment at a point.

So, a beam can fail in shear or it can fail in bending. A natural question at this stage is: which will occur first? Unfortunately, there is no general answer to that question. In some circumstances, a beam will fail in shear; in other cases, a beam will fail in bending. Which happens first depends on the longitudinal profile of the beam: its spans, the position and nature of its supports and the positions and magnitudes of the loading on it. Only by calculation can we tell whether a shear or a bending failure will occur first.

The first thing we need to do is develop a system of *quantifying* shear and bending effects. These quantifications are called *shear force* and *bending moment* respectively and are defined in the following paragraphs.

Shear force

A shear force is the force tending to produce a shear failure at a given point in a beam.

The value of shear force at any point in a beam = the *algebraic sum* of all upward and downward forces to the left of the point. (The term 'algebraic sum' means that upward forces are regarded as being positive and downward forces are considered to be negative.)

Example 16.1
Consider the example shown in Fig. 16.4, in which the end reactions have already been calculated as 25 kN and 15 kN as shown (you should check this). To calculate the shear force at point A, ignore everything to the right of A and examine all the forces that exist to the left of A. Remember, upward forces are positive and downward forces are negative. Adding the forces together:

Shear force at A $= +25 - 30 - 10 = -15\,$kN

Bending moment

The bending moment is the magnitude of the bending effect at any point in a beam. We encountered moments in Chapter 8, where we learned that a moment is a force multiplied by a perpendicular distance, it's either clockwise or anticlockwise and is

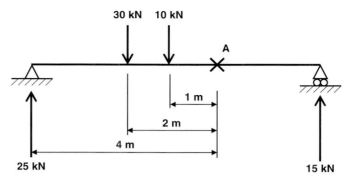

Fig. 16.5   Bending moment at point A.

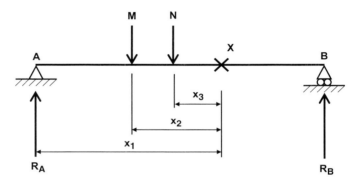

Fig. 16.6   Shear forces and bending moments – general case.

measured in kN.m or N.mm. The value of bending moment at any point on a beam = the sum of all bending moments to the left of the point. (Regard clockwise moments as being positive and anticlockwise moments as being negative.)

Consider again the beam shown in Fig. 16.4. To calculate the bending moment at point A, ignore everything to the right of A and examine the forces (and hence moments) that exist to the left of A. You should realise that, as we are calculating the moment at A, all distances should be measured from point A to the position of the relevant force. See Fig. 16.5 for clarification.

$$\text{Bending moment at A} = (25\,\text{kN} \times 4\,\text{m}) - (30\,\text{kN} \times 2\,\text{m}) - (10\,\text{kN} \times 1\,\text{m})$$
$$= 100 - 60 - 10$$
$$= 30\,\text{kN.m}$$

Figure 16.6 shows a more generalised case. Beam AB supports two point loads, M and N, located at the positions shown. The end reactions at A and B are R_A and R_B respectively. Suppose that we are interested in finding the shear force at position X, which is located a distance x_1 from the support A, x_2 from point load M and x_3 from point load N. The shear force and bending moment at X are calculated as follows:

Shear force at $X = R_A - M - N$

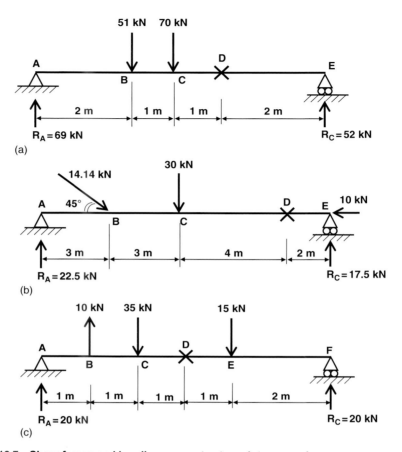

Fig. 16.7   Shear forces and bending moments at a point – examples.

Bending moment at $X = (R_A \times x_1) - (M \times x_2) - (N \times x_3)$

(Remember: clockwise moments are positive, anticlockwise moments are negative.)

Shear force and bending moment: some examples

In each of the three examples shown in Fig. 16.7, calculate the shear force and bending moment at point D. Check your answers with those given below:

a)   Shear force at $D = -52\,kN$; bending moment at $D = 104\,kN.m$.
b)   Shear force at $D = -17.5\,kN$; bending moment at $D = 35\,kN.m$.
c)   Shear force at $D = -5\,kN$; bending moment at $D = 45\,kN.m$.

(If you are unsure where these answers came from, reread the examples and rules given above. In example b), the vertical component of the inclined 14.14 kN force is 10 kN; revisit Chapter 7 for clarification.)

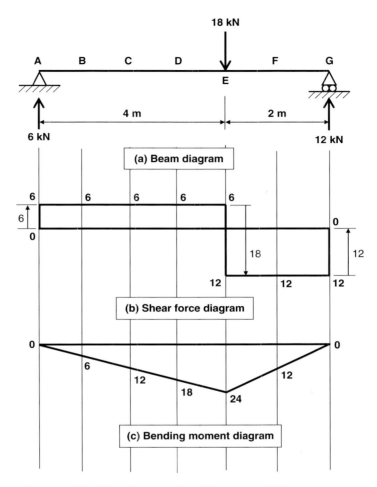

Fig. 16.8   Example 16.2 – Shear force and bending moment diagrams.

Up till now we've discussed how to calculate values of shear force and bending moment at a specific point in a beam. As engineers though, we're not interested so much in the values at a specific point as in how shear force and bending moment vary along the entire length of a beam. Accordingly, we can calculate and draw graphical representations of shear force and bending moment and their variation along a beam. These are called shear force and bending moment diagrams.

Shear force and bending moment diagrams

Example 16.2

Look at the example shown in Fig. 16.8a. The beam is supported at its two ends, A and G, and experiences an 18 kN point load at point E, which is 4 metres from the beam's left-hand end. The reactions at the left and right hand ends are 6 kN and 12 kN respectively, as previously calculated in Chapter 9.

We are going to calculate the shear force and bending moment values at 1 metre intervals along the beam, in other words at points A, B, C, D, E, F and G. When you do this or a similar exercise yourself, I suggest you use graph paper and draw vertical guidelines to make the draughtsmanship easier.

Shear forces

(Remember, always look at what's going on to the *left* of the point at which you're trying to calculate shear force.) First of all, draw a horizontal straight line representing zero shear force. This will be the base line from which the shear force diagram is drawn.

There is nothing to the left of point A, so the shear force at point A is zero.

If we go a very small distance (say 2 mm) to the right of A, there is now a 6 kN upward force to the left of the point we're considering. So the shear force at this point is 6 kN. We can represent this effect by a vertical straight line at point A, starting at the zero force base line and going up to a point representing 6 kN. Each of points B, C, D and E has a 6 kN force to the left of it (i.e. the reaction at point A), so the shear force at each of those points is 6 kN. These values can be plotted on our shear force diagram.

Now consider a point a very small distance (say 2 mm) to the right of E. If we examine all the forces to the left of this point, we see that there is an upward force of 6 kN (at A) and a downward force of 18 kN (at E). The shear force at this point must be $(6 - 18) = -12$ kN (which means 12 kN below the base line). The shear forces at F and just to the left of G will have the same value (-12 kN).

At G itself the sum of all the forces $= (6 \text{ kN} - 18 \text{ kN} + 12 \text{ kN}) = 0$ kN. So the shear force at G is zero. The shear force diagram is drawn in Fig. 16.8b.

Bending moments

Again, we will be looking solely at forces and moments to the left of the point we're considering. We will calculate the moment at each point, remembering that:

- clockwise moments are positive and anticlockwise moments are negative
- distances are measured from the force concerned to the point considered.

Bending moment at A $= +(6 \text{ kN} \times 0 \text{ m})$	$= 0$ kN.m
Bending moment at B $= +(6 \text{ kN} \times 1 \text{ m})$	$= 6$ kN.m
Bending moment at C $= +(6 \text{ kN} \times 2 \text{ m})$	$= 12$ kN.m
Bending moment at D $= +(6 \text{ kN} \times 3 \text{ m})$	$= 18$ kN.m
Bending moment at E $= +(6 \text{ kN} \times 4 \text{ m}) - (18 \text{ kN} \times 0 \text{ m})$	$= 24$ kN.m
Bending moment at F $= +(6 \text{ kN} \times 5 \text{ m}) - (18 \text{ kN} \times 1 \text{ m})$	$= 12$ kN.m
Bending moment at G $= +(6 \text{ kN} \times 6 \text{ m}) - (18 \text{ kN} \times 2 \text{ m})$	$= 0$ kN.m

The bending moment diagram is drawn in Fig. 16.8c.

Hint: As we're only looking at shear forces and bending moments to the left of a particular point, you might find it helpful, to begin with, to use a piece of paper to cover up the part of the diagram to the right of the point you're considering.

There is an easier way

While the above example has given us a good feel for the way in which to calculate and construct shear force and bending moment diagrams, considering every metre along the beam in this way does get rather tedious. A quicker way of drawing the shear force and bending moment diagrams for the above example is as follows.

Shear force diagram – 'follow the arrows'

Draw a base line representing zero shear force. Then start from the left-hand end of the beam. At this point, there is an upward force of 6 kN. So draw a line upwards from the zero line – go up 6 kN, to a value of +6 kN. Going right from A, we encounter no further forces or other features until we reach point E, so the shear force diagram between A and E will be represented by a horizontal straight line between these two points at a value of +6 kN.

At point E there is a downward force of 18 kN. Our shear force diagram will reflect this by dropping down by 18 kN, which takes us from +6 kN to −12 kN. Going right from E, we encounter no further forces or other features until we reach point G, so the shear force diagram between E and G is a horizontal straight line at a value of −12 kN.

At point G there is an upward force of 12 kN. We're already at −12 kN, so the upward force of 12 kN takes us back up to zero. (Note that shear force diagrams *always* end up back on the zero line. If yours doesn't, you've made a mistake somewhere.)

The shear force diagram is shown in Fig. 16.8b. Of course, it is the same as calculated before. Note that there is nothing 'magic' about this process. All we've done is follow the arrows. To summarise: if a force goes upwards (for example, the 6 kN reaction at A), then the shear force diagram goes up by that amount. On the other hand, if a force goes downwards (for example, the 18 kN force at E), then the shear force diagram jumps downwards at that point, again by the same amount.

Bending moment diagram – at 'eventful' points only and join the dots

Earlier we calculated the bending moment at 1 metre intervals along the beam. In fact, we need to do this only at 'eventful' points, plot the values and join the dots. 'Eventful' points (my term) are those points where the problem has some feature, e.g. a point load, a reaction or an end of the beam. If in doubt as to whether a particular point is 'eventful' or not, assume that it is. The 'eventful' points on this beam are A, E and G. We previously calculated the moment values at these three points as 0, 24 and 0 kN.m respectively. Plot these values and join the plotted points with straight lines and you have the bending moment diagram shown in Fig. 16.8c.

There is one further point to note. You may have been wondering why we elected, in Fig. 16.8c, to indicate the bending moment values below the zero line rather than above it. The convention is that the bending moment diagram is plotted on the side of the beam that experiences tension. From the discussion at the beginning of this chapter – and, specifically, from Fig. 16.1 – you will note that in the current example the beam will sag, so tension occurs in the underside of the beam, which suggests that we plot the bending moment diagram below the zero line.

To summarise: the bending moment diagram is drawn either above or below the zero line, depending on whether the beam experiences tension in the top or bottom at the point concerned (top: above the line, bottom: below the line).

The shape of shear force and bending moment diagrams

If you examine the shape of the shear force and bending moment diagrams above you will notice the following features:

- The shear force diagram is a series of 'steps'; in other words, it contains horizontal and vertical straight lines only.
- The bending moment diagram comprises sloping straight lines.

The above features hold for all cases where a beam is loaded with point loads only (i.e. no uniformly distributed loads).

To summarise: if a beam experiences point loads only, the shear force diagram will be a series of steps and the bending moment diagram will contain only straight lines (usually sloping).

The relationship between shear force and bending moment

You may explore the mathematical relationship between shear force and bending moment at a later stage in your course. One thing to be aware of now is the following rule, which always holds:

Where the shear force is zero, the bending moment is either a local maximum, a local minimum or zero.

If we look again at the example in Fig. 16.8, we see that the shear force diagram touches (or cuts through) the zero line at A, E and G. If we look at the bending moment at each of those three points, we see it is zero at A and G and a maximum (24 kN.m) at E.

This rule is very useful in problems where it is difficult to identify the position of maximum bending moment. In such cases, the key lies in identifying the position(s) of zero shear force.

More examples

Draw the shear force and bending moment diagrams for each of the three beams shown in Fig. 16.7. The solutions are given in Figs 16.21–16.23 at the end of this chapter.

Shear force and bending moment diagrams for uniformly distributed loads

In Chapter 9 we saw how to calculate moments for uniformly distributed loads. You might find it worthwhile to revisit that chapter to refresh your memory. The rule for calculating bending moments for uniformly distributed loads is shown in Fig. 9.5 which, for convenience, is reproduced here as Fig. 16.9. With reference to that figure, the moment of the uniformly distributed load about A is the total load multiplied by

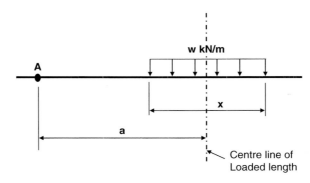

Fig. 16.9   Bending moment calculation for uniformly distributed load (UDL) – general case.

the distance from the centre line of the UDL to the point about which we're taking moments. The total UDL is $w \times x$, the distance concerned is a, so:

Moment of UDL about A $= w \times a \times x$

Apply this principle whenever you're working with uniformly distributed loads.

Example 16.3
Beam AG, shown in Fig. 16.10, spans 6 metres. It supports a uniformly distributed load of 4 kN/m along its entire length. Draw the shear force and bending moment diagrams.

First of all, calculate the reactions. This is easy in this case because of the symmetry of both the beam itself and its loading. Each end reaction will be half the total load on the beam. So

$$R_A = R_G = (4 \, \text{kN/m} \times 6 \, \text{m})/2 = 12 \, \text{kN}$$

We will now try the metre-by-metre approach – as pioneered in the earlier example – to drawing the shear force and bending moment diagrams. So, we are going to calculate the shear force and bending moment values at points A, B, C, D, E, F and G.

Shear forces

(Remember that always look at what's going on to the *left* of the point at which you're trying to calculate shear force.) As before, draw a horizontal straight line representing zero shear force. This will be the base line from which the shear force diagram is drawn.

There is nothing to the left of point A, so the shear force at point A is zero.

If we go a very small distance (say 2 mm) to the right of A, there is now a 12 kN upward force (the reaction at A) to the left of the point we're considering. So the shear force at this point is 12 kN. We can represent this effect by a vertical straight line at point A, starting at the zero force base line and going up to a point representing 12 kN.

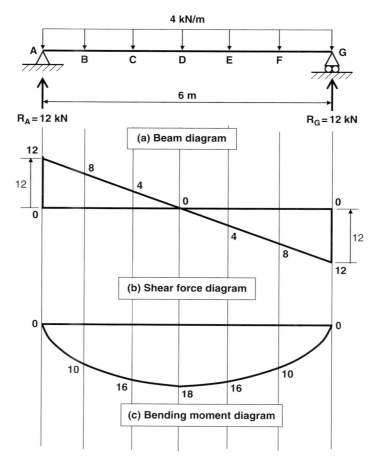

Fig. 16.10   Example 16.3 – Shear force and bending moment diagrams – uniformly distributed load example.

Each of points B, C, D, E, F and G has this 12 kN upward force to the left of it (i.e. the reaction at point A), but they also have downward forces to the left. Let's consider each of these points in turn.

Point B:
Upward force to left = 12 kN.
Downward force to left = $(4\,\text{kN/m} \times 1\,\text{m}) = 4\,\text{kN}$.
Therefore shear force at point B = $12 - 4 = 8\,\text{kN}$.

Point C:
Upward force to left = 12 kN.
Downward force to left = $(4\,\text{kN/m} \times 2\,\text{m}) = 8\,\text{kN}$.
Therefore shear force at point C = $12 - 8 = 4\,\text{kN}$.

Point D:
Upward force to left = 12 kN.
Downward force to left = $(4\,\text{kN/m} \times 3\,\text{m}) = 12\,\text{kN}$.
Therefore shear force at point D = $12 - 12 = 0\,\text{kN}$.

Point E:
Upward force to left = 12 kN.
Downward force to left = (4 kN/m × 4 m) = 16 kN.
Therefore shear force at point E = 12 − 16 = −4 kN.

Point F:
Upward force to left = 12 kN.
Downward force to left = (4 kN/m × 5 m) = 20 kN.
Therefore shear force at point F = 12 − 20 = −8 kN.

Immediately left of Point G:
Upward force to left = 12 kN.
Downward force to left = (4 kN/m × 6 m) = 24 kN.
Therefore shear force left of point G = 12 − 24 = −12 kN.
At point G, there is an upward reaction of 12 kN. So the net shear force at G will be −12 + 12 = 0 kN.

These values can be plotted on our shear force diagram in Fig. 16.10b.

Bending moments

Once more, we will be looking solely at forces and moments to the left of the point we're considering. As in earlier examples, we will calculate the moment at each point, remembering that:

- clockwise moments are positive, and anticlockwise moments are negative
- distances are measured from the force concerned to the point considered.

$$\text{Bending moment at A} = +(12\,\text{kN} \times 0\,\text{m})$$
$$= 0\,\text{kN.m}$$
$$\text{Bending moment at B} = +(12\,\text{kN} \times 1\,\text{m}) - (4\,\text{kN/m} \times 1\,\text{m} \times 0.5\,\text{m}) = 12 - 2$$
$$= 10\,\text{kN.m.}$$
$$\text{Bending moment at C} = +(12\,\text{kN} \times 2\,\text{m}) - (4\,\text{kN/m} \times 2\,\text{m} \times 1\,\text{m}) = 24 - 8$$
$$= 16\,\text{kN.m.}$$
$$\text{Bending moment at D} = +(12\,\text{kN} \times 3\,\text{m}) - (4\,\text{kN/m} \times 3\,\text{m} \times 1.5\,\text{m}) = 36 - 18$$
$$= 18\,\text{kN.m}$$
$$\text{Bending moment at E} = +(12\,\text{kN} \times 4\,\text{m}) - (4\,\text{kN/m} \times 4\,\text{m} \times 2\,\text{m}) = 48 - 32$$
$$= 16\,\text{kN.m}$$
$$\text{Bending moment at F} = +(12\,\text{kN} \times 5\,\text{m}) - (4\,\text{kN/m} \times 5\,\text{m} \times 2.5\,\text{m}) = 0 - 50$$
$$= 10\,\text{kN.m}$$
$$\text{Bending moment at G} = +(12\,\text{kN} \times 6\,\text{m}) - (4\,\text{kN/m} \times 6\,\text{m} \times 3\,\text{m}) = 72 - 72$$
$$= 0\,\text{kN.m}$$

The bending moment diagram is drawn in Fig. 16.10c.

The shape of shear force and bending moment diagrams where uniformly distributed loads are present

If you examine the shape of the shear force and bending moment diagrams in Fig. 16.10 you will notice the following features:

- The shear force diagram comprises sloping straight lines.
- The bending moment diagram is curved (parabolic).

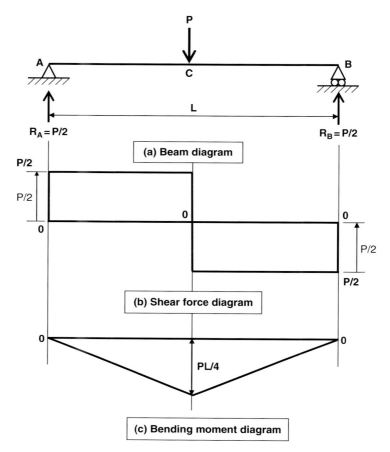

Fig. 16.11   Standard case 1 – Shear force and bending moment diagrams for a beam carrying a central point load.

In general, where a beam is loaded with uniformly distributed loads along all or part of its length, the shear force and bending moment diagrams along the part of the beam concerned have the above features.

To summarise: where a beam experiences uniformly distributed loads, the shear force diagram will comprise sloping straight lines and the bending moment diagram will be curved.

Shear force and bending moment diagrams for standard cases

There are three standard cases of beam loading that are so common that the reader would be well advised to commit the results to memory. These are:

- beam with a central point load
- beam with a non-central point load
- beam carrying a uniformly distributed load over its entire length.

These cases, along with their respective shear force and bending moment diagrams, are shown in Figs 16.11–16.13. Using the techniques discussed above, you should

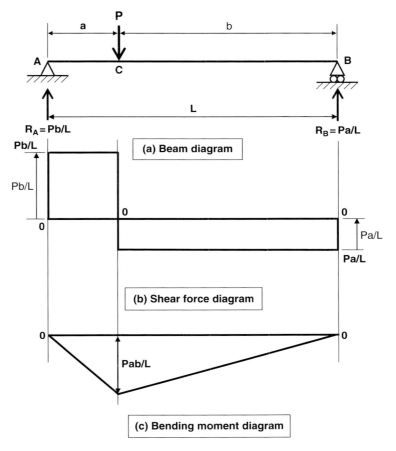

Fig. 16.12   Standard case 2 – Shear force and bending moment diagrams for a beam carrying a non-central point load.

be able to obtain these reactions and shear force and bending moment values for yourself.

Note that the result for the maximum bending moment in a beam with a uniformly distributed load over its entire length ($wL^2/8$) is particularly commonly used in practice – see Fig. 16.13.

Some years ago a colleague of mine in a firm of consulting engineers declared, slightly flippantly: '$wL^2/8$ – that's all you ever need to know!' While this is not quite true (or fair), the comment does at least demonstrate the importance of this result. You might consider that this is underlined by the fact that, in a 'friendly' rafting competition held between the contractor's and resident engineer's staff at a site on which I once worked, the winning raft had been named 'Double You Ell Squared Upon Eight'.

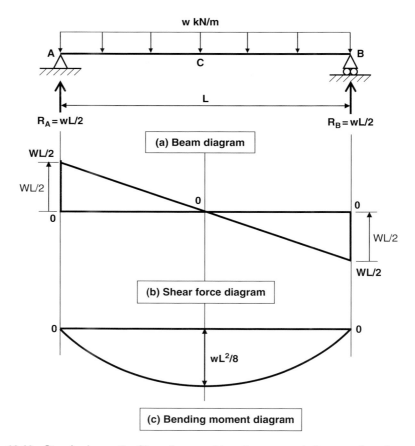

Fig. 16.13   Standard case 3 – Shear force and bending moment diagrams for a beam carrying a uniformly distributed load over its entire length.

An example involving both point loads and uniformly distributed loads

Now we've looked at some examples involving point loads and some involving uniformly distributed loads, as well as some standard cases, let's look at a non-standard example involving both.

As you can see, the example shown in Figure 16.14a contains a number of point loads as well as two uniformly distributed loads (UDLs) of different intensities. This might look a bit challenging at first glance, but if we follow the rules we can easily determine the reactions, and hence the shear force and bending moment diagrams for this case.

As always, we will start by calculating the reactions. First of all we will calculate the total downward load.

$$\begin{aligned} \text{Total downward load} &= 30\,\text{kN} + (5\,\text{kN/m} \times 2\,\text{m}) + 14\,\text{kN} + 10\,\text{kN} \\ &\quad + (10\,\text{kN/m} \times 3\,\text{m}) \\ &= 94\,\text{kN}. \end{aligned}$$

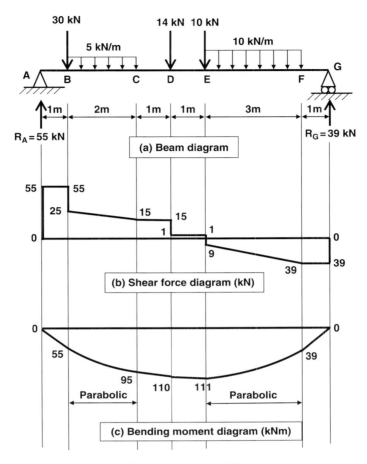

Fig. 16.14   Example involving both point loads and UDLs.

For equilibrium, total force up = total force down, so the sum of the two reactions must be 94 kN.

$$R_A + R_G = 94\,\text{kN}.$$

Taking moments about A:
Total clockwise moment = total anticlockwise moment.

$$(30\,\text{kN} \times 1\,\text{m}) + (5\,\text{kN/m} \times 2\,\text{m} \times 2\,\text{m}) + (14\,\text{kN} \times 4\,\text{m}) + (10\,\text{kN} \times 5\,\text{m})$$
$$+ (10\,\text{kN} \times 3\,\text{m} \times 6.5\,\text{m}) = (R_G \times 9\,\text{m})$$
$$9R_G = 30 + 20 + 56 + 50 + 195 = 351\,\text{kN}$$
$$R_G = 39\,\text{kN}$$

Taking moments about G:
Total clockwise moment = total anticlockwise moment.

$$(R_A \times 9\,\mathrm{m}) = (30\,\mathrm{kN} \times 8\,\mathrm{m}) + (5\,\mathrm{kN} \times 2\,\mathrm{m} \times 7\,\mathrm{m}) + (14\,\mathrm{kN} \times 5\,\mathrm{m}) + (10\,\mathrm{kN} \times 4\,\mathrm{m}) +$$
$$(10\,\mathrm{kN/m} \times 3\,\mathrm{m} \times 2.5\,\mathrm{m})$$
$$9R_A = 240 + 70 + 70 + 40 + 75 = 495\,\mathrm{kN}$$
$$R_A = 55\,\mathrm{kN}.$$

Check: $R_A + R_G = 55 + 39 = 94\,\mathrm{kN}$ (correct)

Now the reactions have been calculated, the shear force diagram may be drawn. As we saw earlier in this chapter, in the section headed 'There is an easier way', the shear force can be drawn. As in earlier examples, start by drawing a horizontal line representing zero shear force. Then, starting at the left-hand end of the beam (point A) construct the shear force diagram as follows:

- At point A, draw a vertical line up from the base line to a value of 55 kN. This represents the upward reaction, R_A, which has a value of 55 kN.
- As there are no loads between A and B, draw a horizontal straight line rightwards from A.
- At B, draw a vertical line downwards of length 30 kN, which represents the downward force of 30 kN. The value arrived at is 25 kN (i.e. 55 – 30).
- Between B and C there is a udl of 5 kN/m, which means that the shear force 'loses' 5 kN for every metre travelled. The distance between B and C is 2 metres, which suggests a total loss of shear force of (5 × 2) = 10 kN over this length, represented by a sloping straight line. The value arrived at at C is 15 kN (i.e. 25 – 10).
- Between C and D there is no loading, so the shear force value remains at 15 kN. This is represented by a horizontal straight line between these two points.
- At D, draw a vertical line downwards of length 14 kN, which represents the downward force of 14 kN. The value arrived at is 1 kN (i.e. 15 – 14).
- Between D and E there is no loading, so the shear force value remains at 1 kN. This is represented by a horizontal straight line between these two points.
- At E, draw a vertical line downwards of length 10 kN, which represents the downward force of 10 kN. The value arrived at is –9 kN (i.e. 1 – 10).
- Between E and F, there is a UDL of 10 kN/m, which means that the shear force 'loses' 10 kN for every metre travelled. The distance between B and C is 3 metres, which suggests a total loss of shear force of (10 × 3) = 30 kN over this length, represented by a sloping straight line. The value arrived at at C is –39 kN (i.e. –9 – 30).
- Between F and G there is no loading, so the shear force value remains at –39 kN. This is represented by a horizontal straight line between these two points.
- At G, draw a vertical line upwards of length 39 kN, which represents the upward reaction, R_G, of 39 kN. The value arrived at is 0 kN (i.e. – 39 + 39).

As always, the shear force diagram returns exactly to a value of zero at the right-hand end. The shear force diagram is shown in Figure 16.14b.

Now we can turn our attention to the bending moment diagram. Calculate the bending moment at each of points A–G, plot the points on a graph, then join the dots on the graph with either a straight line (if there is no loading between the points) or a curved line (if there is a UDL between the points).

Bending moment at A = 0 kN.m

Bending moment at B = (55 kN × 1 m) = 55 kN.m

Bending moment at C = (55 kN × 3 m) − (30 kN × 2 m) − (5 kN/m × 2 m × 1 m)

    = 95 kN.m

Bending moment at D = (55 kN × 4 m) − (30 kN × 3 m) − (5 kN/m × 2 m × 2 m)

    = 110 kN.m

Bending moment at E = (55 kN × 5 m) − (30 kN × 4 m) − (5 kN/m × 2 m × 3 m)

    − (14 kN × 1 m) = 111 kN.m

Bending moment at F = (55 kN × 8 m) − (30 kN × 7 m) − (5 kN/m × 2 m × 6 m)

    − (14 kN × 4 m) − (10 kN × 3 m) − (10 kN/m × 3 m × 1.5 m)

    = 39 kN.m

Bending moment at G = 0 kN.m

The bending moment diagram is shown in Figure 16.14c.

Note the following:

- There is no loading between points A and B, C and D, D and E and F and G, so the bending moment diagram between these points is a sloping straight line.
- There is a uniformly distributed load (UDL) between B and C, and also between E and F, so the bending moment diagram between these points is a curved (parabolic) line.
- The shear force diagram crosses the zero line at E so, as expected, the maximum bending moment value (111 kN.m) also occurs at E.

More examples involving uniformly distributed loads

Draw the shear force and bending moment diagrams for each of the beams shown in Figs 16.15. The solutions are given in Figs 16.24–16.26 at the end of this chapter.

What else can shear force and bending moment diagrams tell us?

Look at the beam shown in Fig. 16.16a. It is supported at A and C and experiences a point load at B and at the free end D. By examining the beam and deducing the way in which it might bend (in the same way as we did with the examples at the very beginning of this chapter), we can deduce that:

- the beam is sagging at point B
- the beam is hogging at support C
- the beam is hogging at point D.

Clearly, somewhere between points B and C, the nature of the beam's deflection switches from sagging to hogging. This point is termed the ***point of contraflexure***. But where, exactly, does the point of contraflexure occur?

By now you should be able to calculate the reactions and draw the shear force and bending moment diagrams. These are shown in Figs 16.16b and c respectively.

Now, earlier in this chapter you were introduced to a convention which stated that the bending moment diagram is always drawn on the tension side of the zero line. This suggests that:

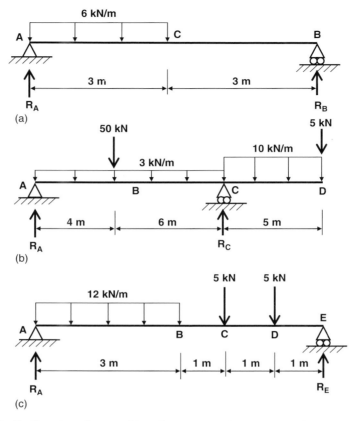

Fig. 16.15   Further shear force and bending moment diagram examples.

- if the bending moment profile is below the zero line, tension occurs in the bottom face of the beam, which suggests it is *sagging*
- if the bending moment profile is above the zero line, tension occurs in the top face of the beam, which suggests it is *hogging*.

It follows from this that where the bending moment diagram crosses the zero line, the nature of deflection of the beam switches from sagging to hogging (or vice versa). Therefore a point of contraflexure occurs wherever the bending moment profile crosses the zero line. In the current example, that point is 2.5 metres from the left-hand end of the beam. This is determined by recognising that the two (hatched) triangles that constitute the bending moment diagram are **similar** (in the mathematical sense of the word). The deflected profile of the beam is shown in Fig. 16.16d.

Example 16.4
Draw the shear force and bending moment diagrams and sketch the deflected form for the beam shown in Fig. 16.17. Identify the position of the points of contraflexure. (The solution is given in Fig. 16.27 at the end of this chapter.)

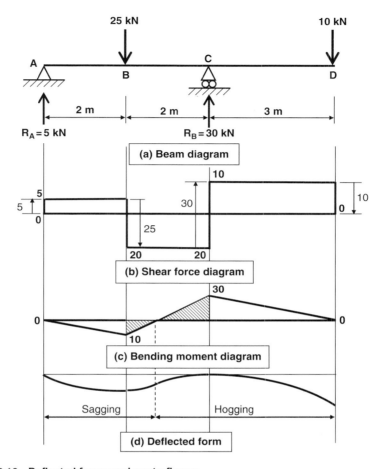

Fig. 16.16   Deflected forms and contraflexure.

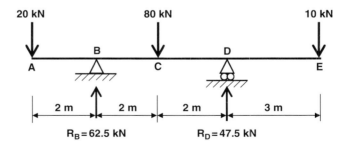

Fig. 16.17   Example 16.4.

More about the point of contraflexure

As mentioned above, the point of contraflexure is where the mode of bending of a beam switches from hogging to sagging (or vice versa) and therefore the stresses in

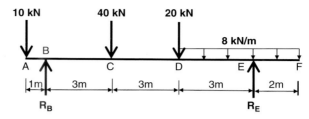

Fig. 16.18   Example 16.5.

the top and bottom outer fibres switch from tension to compression (or vice versa). Its position is where the bending moment diagram crosses the zero line.

The position of the point of contraflexure is particularly significant in concrete beams, which are weak in tension and therefore are reinforced with steel bars. The point of contraflexure identifies where the tension zone stops – or starts – in any particular face of the beam (i.e. top or bottom) and therefore the point at which reinforcement is required.

Let's look at another example.

Example 16.5

Figure 16.18 shows a beam loaded with point loads at A, C and D, and a uniformly distributed load between points D and F. The beam is supported at points B and E. Calculate the support reactions R_B and R_E, draw the shear force and bending moment diagrams, and identify the position of the points of contraflexure.

Solution

See Figure 16.19 for diagrams. The calculations are below.

Total downward load $= 10 + 40 + 20 + (8\,\text{kN/m} \times 5\,\text{m}) = 110\,\text{kN}$.

Taking moments about B:

$(40\,\text{kN} \times 3\,\text{m}) + (20\,\text{kN} \times 6\,\text{m}) + (8\,\text{kN} \times 5\,\text{m} \times 8.5\,\text{m}) = 9R_E + (10\,\text{kN} \times 1\,\text{m})$.
$9R_E = 120 + 120 + 340 - 10 = 570\,\text{kN}$
$R_E = 570/9 = 63.33\,\text{kN}$
$R_E = 63.33\,\text{kN}$

Taking moments about E:

$9R_B = (10\,\text{kN} \times 10\,\text{m}) + (40\,\text{kN} \times 6\,\text{m}) + (20\,\text{kN} \times 3\,\text{m}) + (8\,\text{kN} \times 5\,\text{m} \times 0.5\,\text{m})$.
$9R_B = 100 + 240 + 60 + 20 = 420\,\text{kN}$
$R_B = 420/9 = 46.66\,\text{kN}$
$R_B = 46.66\,\text{kN}$

Check: $R_B + R_E = 46.66 + 63.33 = 110\,\text{kN}$ (correct)

Bending moment at A $= 0$ (by inspection)
Bending moment at B $= -(10\,\text{kN} \times 1\,\text{m}) = -10\,\text{kN.m}$
Bending moment at C $= -(10\,\text{kN} \times 4\,\text{m}) + (46.66\,\text{kN} \times 3\,\text{m}) = 100\,\text{kN.m}$

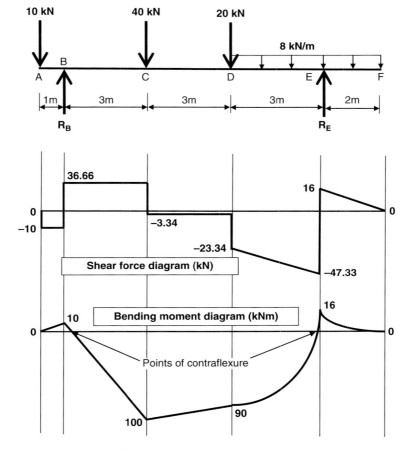

Fig. 16.19   Example 16.5 – Solution

Bending moment at D $= -(10\,kN \times 7\,m) - (40\,kN \times 3\,m) + (46.66\,kN \times 6\,m)$
$= 90\,kN.m$
Bending moment at E $= -(10\,kN \times 10\,m) - (40\,kN \times 6\,m) - (20\,kN \times 3\,m)$
$\qquad - (8\,kN \times 3\,m \times 1.5\,m) + (46.66\,kN \times 9\,m) = -16\,kN.m$
Bending moment at F $= 0$ (by inspection)
Max hogging moment $= 16\,kNm$

Points of contraflexure are indicated on bending moment diagram above.
Position of left-hand point of contraflexure is $(10\,kN \times 3\,m/110\,kN) = 0.27\,m$ right
of B, which is 1.27 m to the right of A.

What you should remember from this chapter

- Shear is a cutting or slicing action which causes a beam to break or snap.
- If a beam is subjected to a load, it will bend. If the loading is increased, the bending
  will increase and eventually the beam will break (if it doesn't fail in shear first).

- A shear force is the force tending to produce a shear failure at a given point in a beam.
- The value of shear force at any point in a beam = the algebraic sum of all upward and downward forces to the left of the point.
- A beam will fail in either bending or shear. Which occurs first can only be determined by calculation.
- The bending moment is the magnitude of the bending effect at any point in a beam. The value of bending moment at any point on a beam = the sum of all bending moments to the left of the point.
- Shear force and bending moment diagrams are graphical representations of shear force and bending moment and their variation along a beam.
- The bending moment diagram is drawn either above or below the zero line, depending on whether the beam experiences tension in the top or bottom at the point concerned (top: above the line, bottom: below the line).
- Where the shear force is zero, the bending moment is either a local maximum, a local minimum or zero. It follows from this that the position of maximum bending moment can be determined from drawing the shear force diagram first.
- If a beam experiences point loads only, the shear force diagram will be a series of steps and the bending moment diagram will contain only straight lines (usually sloping).

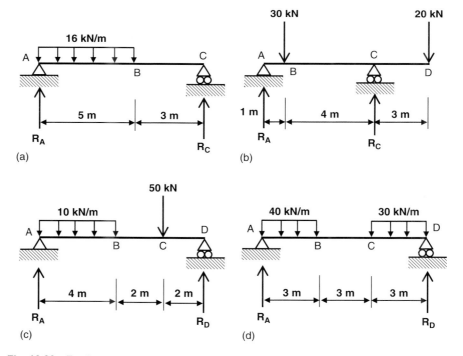

Fig. 16.20   Further tutorial examples.

- Where a beam experiences uniformly distributed loads, the shear force diagram will comprise sloping straight lines and the bending moment diagram will be curved.
- The point of contraflexure is where the deflected form of a beam switches between hogging and sagging or vice versa. The bending moment diagram will cross the zero line at this point.
- And don't forget $wL^2/8$!

Tutorial examples

Draw shear force and bending moment diagrams for each of the beams shown in Fig. 16.20.

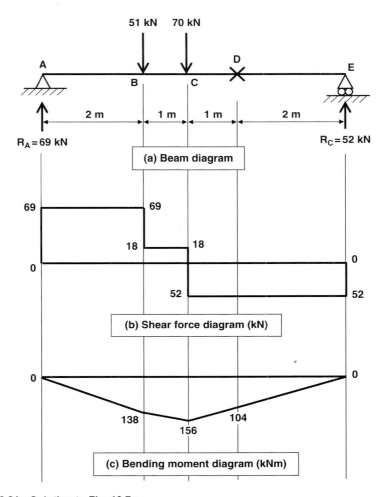

Fig. 16.21   **Solution to Fig. 16.7a.**

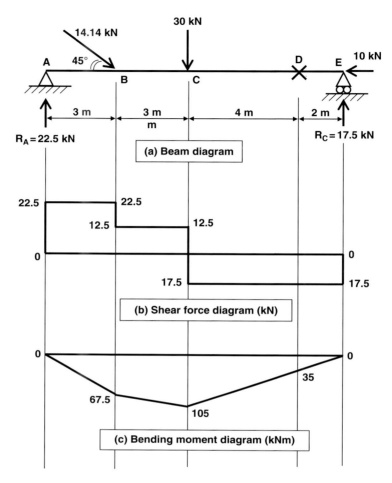

Fig. 16.22   Solution to Fig. 16.7b.

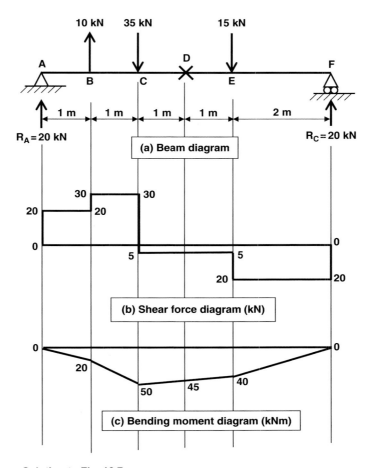

Fig. 16.23   Solution to Fig. 16.7c.

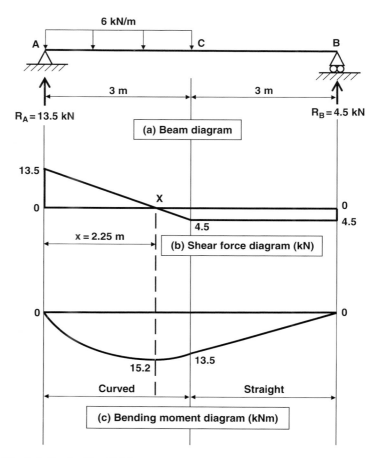

Fig. 16.24   Solution to Fig. 16.14a.

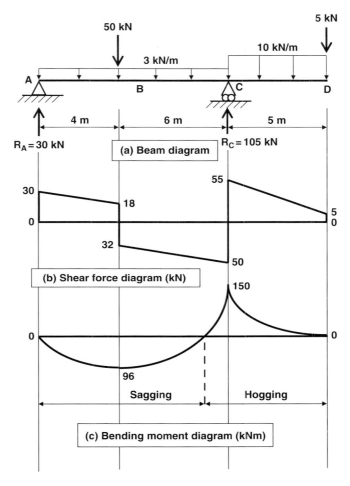

Fig. 16.25   Solution to Fig. 16.14b.

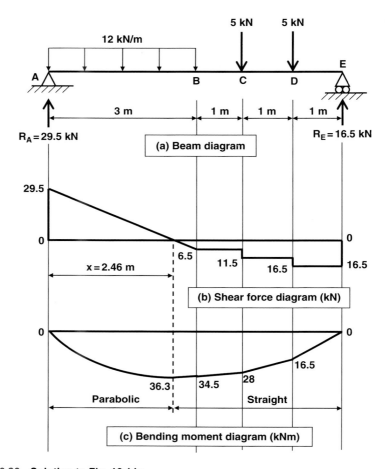

Fig. 16.26   Solution to Fig. 16.14c.

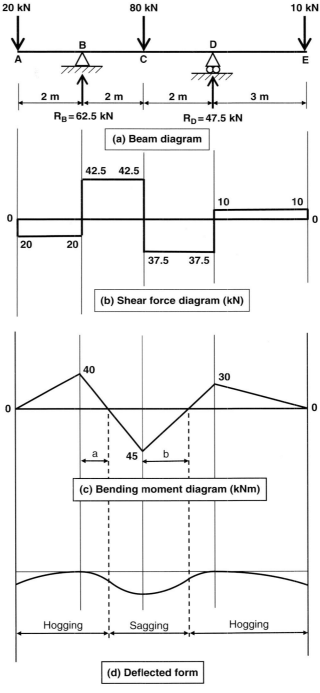

(a) Beam diagram

(b) Shear force diagram (kN)

(c) Bending moment diagram (kNm)

(d) Deflected form

Distance a = 2 m x (40/85) = 0.94 m
Distance b = 2 m x (45/75) = 1.2 m

Fig. 16.27   Solution to Example 16.4.

Fig. 16.28   Cantilevered balcony in shopping centre.

Figure 16.28 shows a cantilever in a shopping centre in Germany. The cantilever allows a cafe area on an upper level to overhang (by a modest distance) the pedestrian circulation area below. Note how the depth of the supporting beam reduces towards the 'free' (i.e. unsupported) end. This is because the bending moment in the beam also reduces towards the free end.

17 This thing called stress

Introduction

If you are studying to become an architect you will be relieved to know that most laymen have some idea of the work that an architect does. However, the civil engineers among you have probably already discovered, to your dismay, that most members of the general public – even the more educated ones – don't have a clue what a civil engineer is or what he or she does, despite the efforts of the relevant professional bodies to promote the profession. However, if pushed, some non-engineers are aware that engineers 'deal with stresses' and that is what this chapter is about.

If some members of the general public are aware of engineers' dealings with stresses, it may have come as a surprise that stresses have hardly been mentioned in the first 16 chapters of this book. However, this and the following three chapters are concerned exclusively with stress.

As we shall see, stress is internal pressure at a point within a structural element occurring as a result of the loads and moments to which the element is subjected. There is a limit to the amount of stress that any given material can take, so in structural design it is important to check that this stress is not exceeded.

Figure 17.1 shows a high-tech tent structure: the roof of the Sony Centre in Berlin.

Fifteen giant steel arches, approximately 7 metres apart, form the main structure of the unusual building shown in Fig. 17.2. The vertical supports to the floors in this building are either suspended off the higher reaches of the arch or, in the case of the end columns, supported off the lower part of the arch. The end support (visible beneath the foliage of the tree in the photograph) is thus in compression and is noticeably fatter than the other vertical supports (which are in tension); this is because it has to be designed against the possibility of buckling (that is, bending and crumpling).

What is stress?

Stress is internal pressure. Pressure is defined mathematically as force/area. As an example, let's suppose you are considering an extended period of foreign travel. As you will be travelling for many months, your extensive preparations will include purchasing a backpack. When fully loaded, the backpack will no doubt be quite heavy, so it's important to select one that will be as comfortable as possible to wear.

Basic Structures, Second Edition. Philip Garrison.
© 2011 John Wiley & Sons, Ltd. Published 2011 by John Wiley & Sons, Ltd.

Fig. 17.1   Roof of Sony Centre, Berlin.

Fig. 17.2   Ludwig-Erhard-Haus ('The Armadillo'), Berlin.

From experience, you know that a backpack with narrow shoulder straps will become uncomfortable – if not downright painful – very quickly. An extremely narrow strap – for example, a piece of string – would become extremely uncomfortable and you would probably whimper in agony as the string cut into your shoulders. This is because the load contained in the backpack is transmitted to your shoulder through a comparatively small area, so the pressure will be large (since pressure = force/(small) area).

Fig. 17.3   **Cable-stayed bridge, Bingley bypass, West Yorkshire.**

On the other hand, a broad-strapped backpack will feel much more comfortable. This is because the load from the contents of the backpack will be spread over a much greater area, hence the pressure will be much less. So the message is: choose a backpack with broad shoulder straps and you'll feel much more comfortable.

Whereas pressure is external to an object – e.g. the pressure transmitted into your shoulder through the straps of a backpack or the pressure on a concrete slab due to a heavy piece of machinery or the pressure a building exerts through its foundations to the ground beneath – stress is a similar phenomenon but considered at a point within (for example) a concrete column, a steel beam or a timber joist.

As for pressure, *direct* stress is defined mathematically as force/area. (Readers should note that it is only direct stress that is thus defined; bending stress and shear stress are different, as we shall see in the chapters that follow.)

Units of stress

As we know, force is measured in newtons (N) or kilonewtons (kN) and area is measured in square millimetres (mm^2) or square metres (m^2). As direct stress is force/area, it could be expressed in units of kN/m^2 or N/mm^2. In civil engineering we use N/mm^2 as the units of stress, for the reason that the stresses encountered in practice can be expressed in manageable figures in N/mm^2 units.

There is a limit to the stress that any particular material can take. This stress is known as the *permissible stress* or the *strength* of the material. Obviously, some materials are stronger than others. For example, the strength of timber is typically in the range 4–7 N/mm^2, depending on the species. The strength of concrete is typically in the range 25–40 N/mm^2, while the strength of the steel type normally used in structural steelwork construction is 275 N/mm^2.

Note the inclination of the main mast of the cable-stayed bridge shown in Fig. 17.3. What does this tell you about the nature of the stresses in the bridge?

Stress and strain

We use the terms stress and strain in everyday life in circumstances unconnected with structures. For example, you hear people say 'he's under stress' or 'she's feeling the strain'. The 'popular' uses of the words stress and strain are analogous to the technical uses of those words, as we shall see.

Stress can arise as a result of certain situations or circumstances. For example, you might find any of the following situations stressful:

● You are on a plane which is being hijacked.
● Your partner has just announced that s/he is leaving you.
● Your boss tells you he will have to 'let you go'.
● Your car breaks down.

You might react to the stressful situation in a number of ways:

● You might get angry and shout at someone.
● You might burst into tears.
● You might decide that a stiff drink would help.

The ***stress*** is represented by the situation (the hijacked plane, wife leaving you, etc.) and the ***strain*** is represented by your reaction to it (the tears, anger or stiff drink).

It's the same principle in structural engineering. For example, a column in a building experiences stress as a result of the forces on it from the floors and walls that the column is supporting. These forces are trying to compress, or squash, the column – in other words, the forces are inflicting ***stress*** on the column. The column will react to this 'squashing' stress by allowing itself to be reduced in length. This reduction in length (as a proportion of the column's original length) is the ***strain***.

Similarly, a hanger cable in a suspension bridge experiences a stress that is trying to stretch the cable, to which it responds by increasing its length. This increase in length (as a proportion of the cable's original length) is the strain.

You will learn more about stress and strain in structural engineering and how to calculate their values in the next chapter.

The graceful shallow concrete arches shown in Fig. 17.4 provide an uncluttered public space beneath an elevated roadway.

Types of stress

Direct and shear stresses are discussed in Chapter 18. Bending stress is explained in Chapter 19. Combined bending and axial stresses are investigated in Chapter 20.

Fig. 17.4   Concrete arches supporting elevated roadway, Lille, France.

What you should remember from this chapter

- Stress is the internal pressure occurring at a given point within a structural element.
- The units of stress are N/mm^2.
- Strain is a measure of what happens as the result of the stress. For example, an extension or reduction in length.

18 Direct (and shear) stress

Introduction

Chapter 17 introduced the concepts of stress and strain. In this chapter we shall discuss direct and shear stresses. We shall also look at how to calculate strains.

Direct (axial) stress

As discussed in Chapter 17, stress is an internal pressure. A direct (axial) stress occurs as a result of a direct (axial) force which acts along the axis of the member, and perpendicular to the member's cross-section. Depending on the direction of the force, the member may experience tension causing extension, or stretching, of the member, or compression which causes contraction, or squashing, of the member. Remember that for equilibrium, forces in one direction must be opposed by equal forces in the opposite direction (see Chapter 6). Examples include a concrete column experiencing a vertical load, as shown in Fig. 18.1a and a steel bar experiencing a horizontal load, shown in Fig. 18.1b.

You will notice that the load in the column in Fig. 18.1a is attempting to squash the column, therefore it is inducing a compressive stress in the column. On the other hand, the force on the steel bar in Fig. 18.1b is trying to stretch the bar, so it is producing a tensile stress in the bar. In both cases, if the values of the force (P) and the cross-sectional area (A) are known, the direct (axial) stress can be calculated using the following equation:

$$\text{Direct stress } (\sigma) = \frac{\text{Force } (P)}{\text{Area } (A)}$$

As explained in Chapter 17, the stress calculated should be expressed in units of N/mm².

It is important to note that the stress has the same value at every point in the cross-section of the column or bar and it is generally assumed that the stress will be the same throughout the length of the element as well.

Shear stress

A shear stress occurs as a result of a shear force. You will remember from earlier chapters that shear is a cutting or slicing action – for example, if you cut through a loaf of bread with a bread knife you are applying a shear force to the loaf. Shear forces therefore

Basic Structures, Second Edition. Philip Garrison.
© 2011 John Wiley & Sons, Ltd. Published 2011 by John Wiley & Sons, Ltd.

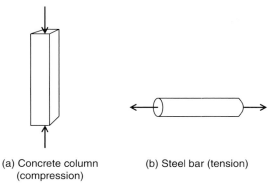

(a) Concrete column
(compression)

(b) Steel bar (tension)

Fig. 18.1   Direct stresses.

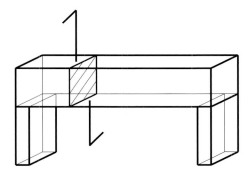

Fig. 18.2   Shear stress in a beam.

act perpendicular to the axis of the member. As with direct stresses, shear forces must be opposed by equal forces in the opposite direction – for example, you wouldn't be able to slice a piece of bread without holding the loaf in place with your other hand (which provides the opposing force) at the same time. An example is a timber beam experiencing a shear force – and hence a shear stress – as shown in Fig. 18.2.

If the hatched zone in Fig. 18.2 represents the cross-section (of area A) where shear failure occurs, and the associated shear force is V, the shear stress is calculated from the following equation:

$$\text{Shear stress } (\tau) = \frac{\text{Shear force } (V)}{\text{Area } (A)}$$

As with direct stress, the shear stress calculated should be expressed in units of N/mm^2.

Note the symbols used for direct stress (σ) and shear stress (τ). These are the standard symbols used in structural engineering. A full list of symbols used in this book is given in Appendix 4.

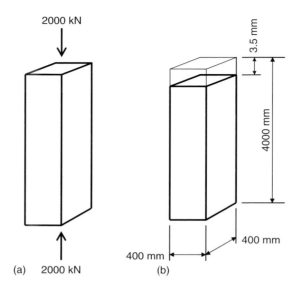

Fig. 18.3   Compressive stress and strain (Example 18.1).

Strain

The concrete column shown in Fig. 18.1a will reduce in length (by a very small amount, it is hoped) as a result of the compressive axial force to which it is subjected. Similarly, the steel bar in Fig. 18.1b will increase in length (again, by a small amount). These changes in length, as a proportion of the original length of the element, give rise to the ***strain***, as defined below:

$$\text{Strain } (\varepsilon) = \frac{\text{Change in length } (\delta L)}{\text{Original length } (L)}$$

It should be pointed out that strain, being simply the ratio of two lengths, has no units. It is a proportion or can, if desired, be expressed as a percentage.

Shear strain is beyond the scope of this book.

We shall now try some numerical examples.

Example 18.1: Stress and strain in compression

A square concrete column in an office building is shown in Fig. 18.3. The column has cross-sectional dimensions 400 mm × 400 mm and supports a total vertical load of 2000 kN. Calculate the direct compressive *stress* at any point in the column.

If the column reduces in length by 3.5 mm as a result of the loading and the column's original length was 4 metres, calculate the *strain* in the column.

Solution

The column is clearly in compression.

The column's cross-sectional area, $A = 400 \times 400 = 160{,}000 \, \text{mm}^2$

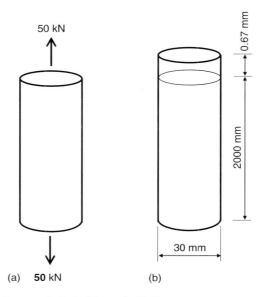

50 kN

0.67 mm

2000 mm

30 mm

(a) **50 kN** (b)

Fig. 18.4 **Tensile stress and strain (Example 18.2).**

Axial load $P = 2000\,\text{kN} = 2000 \times 10^3\,\text{N}$

$$\text{Stress }(\sigma) = \frac{\text{Force }(P)}{\text{Area }(A)} = \frac{2000 \times 10^3}{160,000} = 12.5\,\text{N/mm}^2$$

$$\text{Strain }(\varepsilon) = \frac{\text{Change in length }(\delta L)}{\text{Original length }(L)} = \frac{3.5\,\text{mm}}{4000\,\text{mm}} = 8.75 \times 10^{-4} = 0.000875$$

(Remember: strain has no units.)

Example 18.2: Stress and strain in tension
The circular steel bar shown in Fig. 18.4 has a diameter of 30 mm and is subjected to a tensile axial force of 50 kN. Calculate the direct tensile *stress* at any point in the bar.

If the bar, whose original length was 2 metres, extends in length by 0.67 mm as a result of the force, calculate the *strain* in the bar.

Solution
The procedure is similar to Example 18.1, but this time the member is in tension.

The column's cross-sectional area, $A = \pi r^2 = \pi \times 15^2 = 706.9\,\text{mm}^2$. (Remember that the radius of a circle is half the diameter. The diameter in this case is 30 mm, so the radius is 15 mm.)

Axial load $P = 50\,\text{kN} = 50 \times 10^3\,\text{N}$

$$\text{Stress }(\sigma) = \frac{\text{Force }(P)}{\text{Area }(A)} = \frac{50 \times 10^3\,\text{N}}{706.9\,\text{mm}^2} = 70.73\,\text{N/mm}^2$$

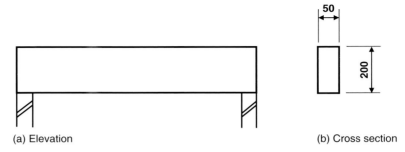

(a) Elevation   (b) Cross section

Fig. 18.5   Timber beam in shear.

$$\text{Strain } (\varepsilon) = \frac{\text{Change in length } (\delta L)}{\text{Original length } (L)} = \frac{0.67 \text{ mm}}{2000 \text{ mm}} = 0.000335$$

Again, strain has no units.

Example 18.3: Shear stress
The shear force at the end of the timber joist shown in Fig. 18.5 is found to be 18 kN. If the timber joist is 50 mm wide and 200 mm deep, calculate the *shear stress* at this point in the joist.

Solution

$$\text{Shear stress } (\tau) = \frac{\text{Shear force } (V)}{\text{Area } (A)} = \frac{18 \times 10^3 \text{ N}}{50 \times 200 \text{ mm}^2} = 1.8 \text{ N/mm}^2$$

The relationship between stress and strain

It would be natural at this point to wonder whether or not there is any relationship between stress and strain. We saw in Chapter 17 that strain is a reaction to stress. In Example 18.1 above, we saw that a stress of 12.5 N/mm^2 in a given concrete column gave rise to a strain of 0.000875. You may wonder whether a stress of double that amount would produce double the strain – or whether tripling the stress would produce triple the strain, and so on. In other words, is stress **proportional** to strain?

If you have studied a materials module you will already know the answer. For most materials, the answer is yes: stress and strain are proportional – up to a point. As you can see from Fig. 18.6, if a graph is plotted of stress versus strain, the graph is a straight line up to a certain point, known as the limit of proportionality. (Beyond the limit of proportionality, the shape of the graph depends on the material but is no longer a straight line.)

If stress is proportional to strain then, mathematically speaking, stress/strain = a constant (Hooke's law). This constant is known as Young's modulus, has the symbol E and units of N/mm^2 or kN/mm^2.

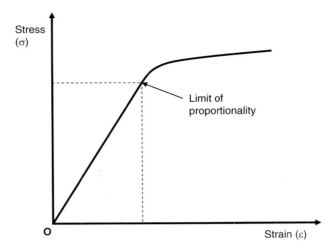

Fig. 18.6   **Stress v. strain graph.**

$$\text{Young's modulus (E)} = \frac{\text{Stress } (\sigma)}{\text{Strain } (\varepsilon)}$$

(For more information on Hooke and Young, see the end of this chapter.)

In Example 18.1 we found that the compressive stress and strain experienced by the concrete column were 12.5 N/mm² and 0.000875 respectively. Therefore:

$$\text{Young's modulus} = \frac{12.5 \text{ N/mm}^2}{0.000875} = 14{,}286 \text{ N / mm}^2 = 14.3 \text{ kN / mm}^2$$

In Example 18.2 we found that the tensile stress and strain experienced by the steel bar were 70.73 N/mm² and 0.000335 respectively. Therefore:

$$\text{Young's modulus for steel} = \frac{70.73 \text{ N/mm}^2}{0.000335} = 211{,}134 \text{ N / mm}^2$$
$$= 211 \text{ kN / mm}^2$$

How to predict change in length

Now we already know that:

$$\text{Direct stress } (\sigma) = \frac{\text{Force } (P)}{\text{Area } (A)}$$

and

$$\text{Strain } (\varepsilon) = \frac{\text{Change in length } (\delta L)}{\text{Original length } (L)}$$

Combining these two equations and rearranging, we get:

$$\text{Change in length } (\delta L) = \frac{PL}{AE}$$

From this equation we can calculate the change in length of a structural element if we know its length (L), the axial load to which it is subjected (P), its cross-sectional area (A) and its Young's modulus value. The latter can be obtained from scientific data tables if necessary.

Example 18.4: Calculating the change in length of a member under direct stress
A steel tie in a space frame roof structure is originally 2 metres long. If the tie is a solid bar of diameter 40 mm, calculate the extension of the steel bar that would be expected if a tensile force of 150 kN is applied to the bar. The Young's modulus of steel is 205 kN/mm². If the extension was unacceptably large, what steps could you take to reduce it?

Solution
Cross-sectional area of steel bar $= \pi r^2 = \pi \times 20^2 = 1256.6 \, \text{mm}^2$.

$$\text{Change in length } (\delta L) = \frac{PL}{AE} = \frac{150 \times 10^3 \, \text{N} \times 2000 \, \text{mm}}{1256.6 \, \text{mm}^2 \times 205 \times 10^3 \, \text{N/mm}^2}$$

$$= 1.16 \, \text{mm}$$

This extension of 1.16 mm is small and is probably tolerable in most structures. However, by examination of the 'change in length' formula, the following steps could be taken to reduce the extension if desired:

- Reduce the axial load in the member.
- Reduce the length of the member.
- Increase the cross-sectional area of the member (this is usually the most practical option).
- Use a material with a greater Young's modulus.

The relationship between change in length and change in width

We have seen that if we apply a compressive force to a structural element such as a column it would shorten, and if a tensile force is applied to a member such as a steel bar it would lengthen. Moreover, we have just seen how to calculate that reduction or increase in length in millimetres by using a simple formula.

Now let us consider what would happen to the *lateral* dimensions when axial forces are applied. If you were to take a piece of plasticine and lengthen it by pulling, you would notice that the girth, or thickness, of the plasticine would get progressively smaller (it would get thinner) until it eventually broke. Shortening is not so easy to demonstrate with plasticine, but if you took a rigid but compressible material and

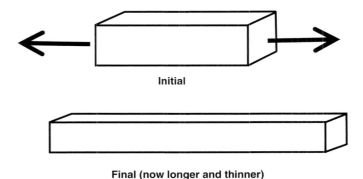

Initial

Final (now longer and thinner)

(a)   Steel bar in tension

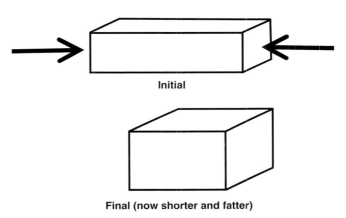

Initial

Final (now shorter and fatter)

(b)   Steel bar in compression

Fig. 18.7   **Steel bars in tension or compression.**

compressed, or squashed, it lengthways you would find that the girth, or thickness, would increase (it would get fatter). See Fig. 18.7.

To recap:

- *Tension* leads to an *increase in length* and a *reduction in width*.
- *Compression* leads to a *reduction in length* and an *increase in width*.

So how do we calculate these changes in width?

How to calculate change in width

As you might suspect, the relationship between the change in length and the change in width is a material property. Experimental work shows that the strain related to change in length (i.e. the longitudinal strain, (ε) is proportional to the lateral strain (ε_L), in the same way that stress is proportional to strain. Earlier in the chapter we saw that the ratio between stress (σ) and strain (ε) is a material property called Young's modulus (E). Perhaps unsurprisingly, the ratio between longitudinal strain (ε) and lateral strain (ε_L) is another material property; this one is called **Poisson's ratio**. Poisson's ratio has the symbol v.

Lateral strain (ε_L)/longitudinal strain (ε) = −Poisson's Ratio (v)

So $\varepsilon_L = -v.\varepsilon$

The minus (−) sign demonstrates that when the lateral strain increases, the longitudinal strain decreases, and vice versa.

Example 18.5: Lateral strain

Calculate the change in lateral dimension, in millimetres, for the steel bar shown in Example 18.4, if the Poisson's ratio for steel is $v = 0.33$.

Longitudinal strain ε = change in length (δL)/original length (L)
$$= 1.16\,\text{mm}/2000\,\text{mm}$$
$$= 0.00058$$

Lateral strain $\varepsilon_L = v.\varepsilon$
$$= 0.33 \times 0.00058$$
$$= 1.914 \times 10^{-4}$$

Change in lateral dimension = Lateral strain × original width
$$= 1.914 \times 10^{-4} \times 40\,\text{mm}$$
$$= 0.0077\,\text{mm}$$

Example 18.6: Lateral strain for an H-shaped cross section

Fig. 18.8 shows a steel column supporting the upper floors of a low-rise office building. The cross-sectional dimensions are as shown. If the axial load on the column is 1500 kN, the column's original length is 3 metres, Young's modulus (E) for steel is 200 kN/mm² and Poisson's ratio (v) is 0.33, calculate:

a) the direct stress (σ) at any point in the column
b) the longitudinal strain (ε) in the column
c) the change in length (δL) of the column
d) the lateral strain (ε_L) in the column
e) the increase in the web width in direction x-x
f) the increase in the web length in direction y-y
g) the increase in the flange width in direction x-x.

Solution

Cross sectional area of column: A = $(250 \times 20) + (250 \times 20) + (210 \times 15)$
$$= 13150\,\text{mm}^2$$
Direct stress $\sigma = P/A = 1500 \times 10^3\,\text{N}/13150\,\text{mm}^2 = 114\,\text{N/mm}^2$
Longitudinal strain $\varepsilon = \sigma/E = 114/200 \times 10^3 = 0.00057$
Change in length $\delta L = \varepsilon.L = 0.00057 \times 3000\,\text{mm} = 1.71\,\text{mm}$
Lateral strain $\varepsilon_L = v.\varepsilon = 0.33 \times 0.00057 = 0.00019$

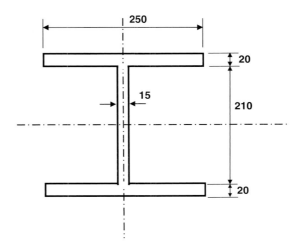

Fig. 18.8   **Example 18.6 (cross section).**

Fig. 18.9   **Grande Arche De La Defense, Paris.**

Increase in the web width in direction x–x = $\varepsilon_L \times 15\,\text{mm} = 0.00285\,\text{mm}$
Increase in the web length in direction y–y = $\varepsilon_L \times 250\,\text{mm} = 0.0475\,\text{mm}$
Increase in the flange width in direction x–x = $\varepsilon_L \times 250\,\text{mm} = 0.0475\,\text{mm}$

Figure 18.9 shows the Grande Arch de la Defense in Paris. Note the tent-like canopy low down within the giant open-sided cube. To give you an idea of scale, Paris's Notre Dame Cathedral could fit beneath this canopy, which is connected to the main arch by a series of steel cables. Using the techniques outlined in the chapter, the designers will have determined the force in each cable and thus calculated the stresses to which the cables are subjected.

What you should remember from this chapter

$$\text{Direct stress } (\sigma) = \frac{\text{Force } (P)}{\text{Area } (A)}$$

$$\text{Shear stress } (\tau) = \frac{\text{Shear force } (V)}{\text{Area } (A)}$$

$$\text{Strain } (\varepsilon) = \frac{\text{Change in length } (\delta L)}{\text{Original length } (L)}$$

$$\text{Young's modulus } (E) = \frac{\text{Stress } (\sigma)}{\text{Strain } (\varepsilon)}$$

$$\text{Change in length } (\delta L) = \frac{PL}{AE}$$

Tutorial examples

1) Calculate the direct stress in a reinforced concrete column of cross section 400 mm × 350 mm, subjected to a compressive load of 3000 kN. Express your answer in N/mm².

2) A solid circular steel rod, forming part of a framed structure, is subjected to a tensile force of 750 kN. If the permissible stress in steel is 460 N/mm², what is the minimum diameter of the rod in millimetres? Had the rod been in compression rather than tension, what other factors would need to be considered?

3) A timber column is subjected to a compressive force of 60 kN. If the permissible compressive stress in timber is 6 N/mm², select a suitable section size for the column. Express your answer in terms of the column's cross-sectional dimensions, in millimetres.

4) A force is applied to a steel bar, originally 3 metres in length, causing it to extend by 1.5 mm. Calculate the strain (ε) in the bar.

5) A 3.5-metre-long steel tie is subjected to a tensile force of 150 kN. If the bar is round, of diameter 20 mm, and the Young's modulus (E) value for steel is 200 kN/mm², calculate the change in length of the bar.

6) An aluminium strut (a compression member) 1.5 metres long is part of a light-weight framed structure and is subjected to a compressive force of 50 kN. Calculate the strain in the strut and determine its change in length. Assume the area of the cross-sectional area is 220 mm² and Young's modulus = 70 kN/mm².

7) A new suspension bridge in the Far East has one of the longest spans in the world. Each of its main cables is 1 metre in diameter and is designed to sustain an axial tensile force of 13,000 tonnes. Assuming, for simplicity, that each main cable is of solid steel (rather than the collection of many smaller diameter cables that it actually is), calculate the stress in each main cable, in N/mm² units.

Tutorial answers

1) $21.4\,\text{N/mm}^2$
2) $45.6\,\text{mm}$; buckling
3) $100\,\text{mm} \times 100\,\text{mm}$ or $75\,\text{mm} \times 150\,\text{mm}$
4) $\varepsilon = 0.0005$, or 0.05%
5) $8.35\,\text{mm}$
6) $\varepsilon = 0.00325$, $\delta L = 4.87\,\text{mm}$
7) $\sigma = 166\,\text{N/mm}^2$

Who were Mr Hooke and Mr Young?

It's claimed that Robert Hooke (1635–1703) was one of the greatest experimental scientists of the 17th century. Certainly he had wide interests in most branches of science, and collaborated with other well-known scientists of the day, including Isaac Newton, but unfortunately the two men did not get on. Hooke also had an interest in architecture and he assisted Sir Christopher Wren on the rebuilding of London's St Paul's Cathedral after the Great Fire of 1666. Hooke's Law, discussed in this chapter, is the scientific principle for which he is best remembered.

Thomas Young (1773–1829) also had wide-ranging professional interests. As well as being a physicist whose experiments in elasticity led to the modulus that bears his name, Young was medically qualified and researched extensively in the fields of light and optics.

19 Bending stress

Introduction

In Chapter 18 we investigated direct stresses – the stresses caused by direct or axial loads on structural elements. In this chapter we will study **bending** stresses. As the name suggests, these are stresses associated with the bending of a beam or other type of structural member.

Bending theory

Consider the beam shown in Fig. 19.1a, which is simply supported at its two ends. If a central point load is applied to the beam, it will bend to give the profile shown in Fig. 19.1b Alternatively, if the beam shown in Fig. 19.1a is subjected to a longitudinal load which does not act along the line of the beam's central axis, it will again bend, to give the profile shown in Fig. 19.1c.

So, bending can be induced in a beam in one of two ways:

1) loading perpendicular to the beam's longitudinal axis
2) eccentric axial loads

If we were to paint vertical stripes at regular intervals along a simply supported beam before loading it, it would appear as shown in Fig. 19.2a. After the beam has bent under loading, its profile will resemble Fig. 19.2b. You will notice that the stripes in the bent beam shown in Fig. 19.2b are still straight, despite the fact that they are no longer the same distance apart at top as they are at the bottom. This would suggest that although the beam has bent, particular cross sections (as represented by the painted stripes) remain straight and thus have not warped.

Consider the cross section of a rectangular beam, shown in Fig. 19.3a. If the beam bends, we know from our earlier studies that the top part of the beam will be in compression and the bottom part will be in tension. This implies that there must be some level in the cross section that will be the interface between the compression and tension zones. This interface is called the **neutral axis** or **neutral plane** and we shall see that there is no stress at this level.

Figure 19.3b is a simple force diagram. The compression in the top part of the beam is represented by force C. The tension in the bottom part of the beam is

Basic Structures, Second Edition. Philip Garrison.
© 2011 John Wiley & Sons, Ltd. Published 2011 by John Wiley & Sons, Ltd.

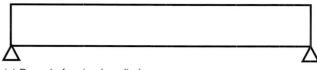

(a) Beam before load applied

(b) Beam bending caused by central point load

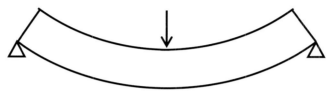

(c) Beam bending caused by eccentric axial load

Fig. 19.1   Bending in beams.

(a) Beam cross sections (edge on) before load applied

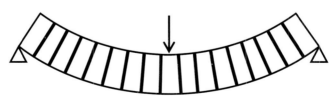

(b) Beam cross sections (edge on) after load applied

Fig. 19.2   Effect of bending on beam cross-section.

represented by force T. Note that, as required for equilibrium, forces C and T are equal but opposite in direction.

Figure 19.3c is a stress diagram in which the vertical line represents zero stress. We can readily see that the maximum tension – and hence the maximum tensile

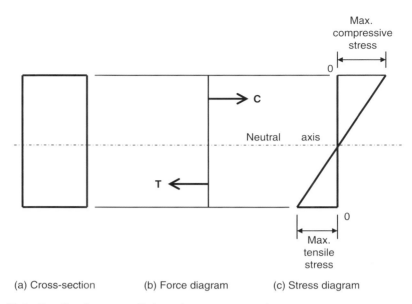

(a) Cross-section   (b) Force diagram   (c) Stress diagram

Fig. 19.3   Bending theory applied to a beam cross-section.

stress – occurs at the very bottom of the beam and reduces as we move up the beam from this level. Similarly, the maximum compression – and hence the maximum compressive stress – occurs at the very top of the beam and reduces as we move down the beam. If we join these two maximum values with a straight line, our stress diagram becomes as shown in Fig. 19.3c. Note the linear (i.e. straight line) variation in stress as we move down the cross section.

As we have just seen (Fig. 19.3c), tension occurs in the bottom of a beam that is sagging. As concrete is weak in tension, steel reinforcement is provided in the place where it is most useful; that is, near the bottom face of the beam. But site labourers in general, and steelfixers in particular, have not been schooled in structural mechanics. Occasionally you will come across cases where a steelfixer feels it is inconvenient to put all the required steel reinforcement in the bottom face and therefore puts some of it half way up a section. Figure 19.3 demonstrates that any steel placed half way up a section is useless, as the stress is minimal at this point and therefore the steel is not doing any work. The bottom of the section is, accordingly, under-reinforced and therefore likely to fail.

Assumptions for bending theory

1) The material is **linearly elastic** (as represented by the straight-line graph in Fig. 19.3c.

2) Young's modulus (E) is the same in compression and tension. (See Chapter 18 if you need a reminder of Young's modulus and its significance.)

3) The material is **homogeneous** (i.e. the same throughout). This is obviously not the case if we're considering a cross section containing two different materials, e.g. reinforced concrete.

4) Plane sections remain plane after bending – i.e. no warping. See the discussion of Fig. 19.2 above.

Neutral axis

As discussed above, the neutral axis occurs at the interface of the compression and tension zones of a structural element experiencing bending. The neutral axis has the following characteristics:

- The neutral axis is the level at which there is no stress.
- The neutral axis is half way down the cross section for homogeneous, symmetrical sections.
- The neutral axis passes through the centroid (see definition below) if the material is homogeneous.

The engineers' bending equation

The equation below is known as the 'engineers' bending equation'. The derivation of it is not included here as it contains some fairly scary mathematics, but it can be found in more advanced structures textbooks. It is far more important for you to become familiar with the equation itself – rather than its derivation – and the meaning of the various terms:

$$\frac{\sigma}{y} = \frac{M}{I} = \frac{E}{R}$$

where:
σ = bending stress (N/mm²)
y = distance (measured, in millimetres, vertically upwards or downwards) to a particular point from the neutral axis (see Fig. 19.4a)
M = bending moment at the point concerned (kN.m or N.mm)
E = Young's Modulus (kN/mm² or N/mm²)
R = radius of curvature (mm) (see Fig. 19.4b)
I = second moment of area (mm⁴) (see explanation below)

The second moment of area is a geometrical property of a cross section. Its derivation is complex, involving calculus. Suffice it to say that for a rectangular section of breadth b and depth d, the second moment of area, $I = bd^3/12$.

Section modulus

A further parameter is the **section modulus**, also known as the elastic modulus. This has the symbol z, and is defined as:

$$z = \frac{I}{y_{max}}$$

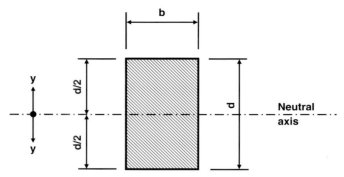

(a) Geometry of a rectangular cross section

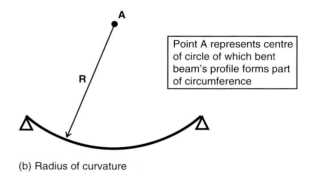

Point A represents centre of circle of which bent beam's profile forms part of circumference

(b) Radius of curvature

Fig. 19.4   Engineers' bending equation – some terms.

Now, for a rectangular section, as mentioned above:

$$I = \frac{bd^3}{12}$$

Moreover, for a rectangular section which is homogeneous (same material through-out), the neutral axis must be exactly half way down the section. Therefore the maximum vertical distance that can be travelled from the neutral axis that still remains within the section is $d/2$. So $y_{max} = d/2$.

So, substituting in the above equation for the special case of a rectangular section:

$$z = \frac{I}{y_{max}} = \frac{\dfrac{bd^3}{12}}{\dfrac{d}{2}} = \frac{bd^3}{12} \times \frac{2}{d} = \frac{bd^2}{6}$$

So, for a rectangular section,

$$z = \frac{bd^2}{6}$$

The basic stress equation

From the engineers' bending equation (discussed above):

$$\frac{\sigma}{y} = \frac{M}{I}$$

Therefore:

$$\sigma = \frac{My}{I}$$

But:

$$z = \frac{I}{y_{max}}$$

Therefore:

$$\sigma = \frac{M}{z}$$

Or, rearranging:

$$z = \frac{M}{\sigma}$$

When I teach this material to students I express the opinion that the above equation is not immediately interesting or exciting, and the reaction I get could be described as passive agreement. However, the above equation – unexciting as it may appear – forms the basis of all structural design. Let me explain.

If a bending moment (M) can be calculated – which it generally can if the loading and span of the beam are known (see Chapter 16) – and the permissible stress (σ) of the material is known (it can be obtained from science data tables), the required section modulus (z) can be determined. Once the required z value is known, a suitable timber beam size or 'off the peg' steel I section modulus can be selected, either by calculation or from tables, as shown in the following two examples.

Example 19.1: Timber beam

A timber beam spans 3.0 metres and carries a uniformly distributed load of 3.35 kN per metre run, as shown in Fig. 19.5. Headroom considerations dictate that a 225 mm deep timber section is used. If the allowable bending stress in timber is 6 N/mm², determine a suitable size (breadth × depth) for the beam.

$$\text{Maximum bending moment } (M) = \frac{wL^2}{8} = \frac{3.35 \times 3^2}{8} = 3.77 \text{ kN.m}$$

$$\sigma = \frac{M}{z} \text{ so, rearranging: } z = \frac{M}{\sigma}$$

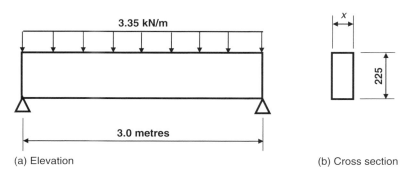

3.35 kN/m

225

3.0 metres

(a) Elevation

(b) Cross section

Fig. 19.5   Sizing a timber beam (Example 19.1).

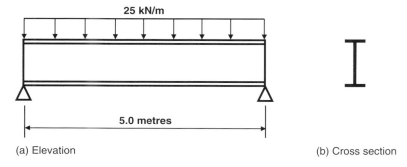

25 kN/m

5.0 metres

(a) Elevation

(b) Cross section

Fig. 19.6   Sizing a steel beam (Example 19.2).

but $z = \dfrac{bd^2}{6}$ for a rectangular section

Therefore $\dfrac{bd^2}{6} = \dfrac{M}{\sigma}$

Rearranging: $b = \dfrac{6M}{\sigma d^2} = \dfrac{6 \times 3.77 \times 10^6 \text{ N.mm}}{6 \text{ N/mm}^2 \times 225^2 \text{ mm}^2}$

So minimum $b = 74.5$ mm

Therefore use a 75 mm wide × 225 mm deep timber beam.

Example 19.2: Steel beam design

A steel beam is to span 5 metres and will carry a load of 25 kN/metre, including its own weight, as shown in Fig. 19.6. If the permissible stress in the steel is 180 N/mm², select a suitable steel beam section from the tables.

From the information above, $w = 25$ kN/m and $L = 5$ m.

Maximum bending moment $(M) = \dfrac{wL^2}{8} = \dfrac{25 \times 5^2}{8} = 78.1$ kN.m

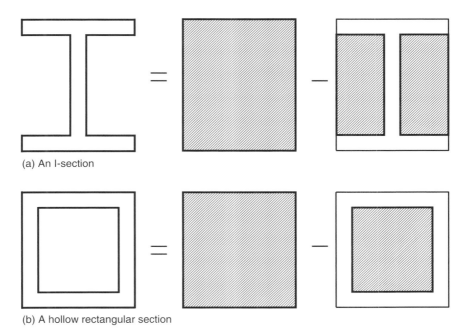

(a) An I-section

(b) A hollow rectangular section

Fig. 19.7   Calculation of second moment of area for common symmetrical shapes.

$$z_{\text{required}} = \frac{M}{\sigma} = \frac{78.1 \times 10^6 \text{ N.mm}}{180 \text{ N/mm}^2} = 433\,889 \text{ mm}^3 = 433.9 \text{ cm}^3$$

Tables of the properties of standard steel beams should now be used. We need to select one that has a section modulus value of 433.9 cm³ or greater. The terminology used in the labelling of steel beams is explained in Chapter 24.

Possibilities include a 305 × 127UB37 steel beam ($z = 471$ cm³) and a 254 × 146UB37 steel beam ($z = 434$ cm³).

If the first of these is selected:

$$\text{Actual bending stress } (\sigma) = \frac{M}{z} = \frac{78.1 \times 10^6}{471\,000 \text{ mm}^3} = 165.8 \text{ N/mm}^2$$

As this is less than the permissible stress of 180 N/mm², this choice is fine. (Note: see Chapter 16 for the origin of $M = wL^2/8$.)

Repeat the above example with a span of 6 metres. You will find that the section modulus (z) value required this time is 625 000 mm³ and therefore a different steel beam section needs to be selected from the tables.

Calculation of second moment of area (*I*) for symmetrical sections

As mentioned above, the *I* value for a rectangular section of breadth b and depth d is $bd^3/12$. Another useful piece of information is that the *I* value for a circle of diameter D is $\pi D^4/64$. Armed with the above information, it is straightforward to calculate *I* values for I sections or hollow rectangular sections (as illustrated in Fig. 19.7) or

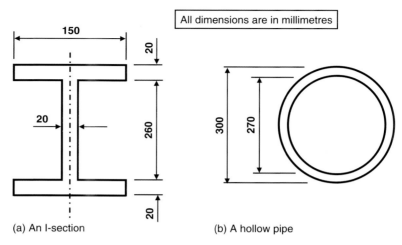

(a) An I-section   (b) A hollow pipe

Fig. 19.8 **Calculate the second moment of area (*I*) value for these symmetrical shapes.**

hollow circular sections. In each case, the shape can be considered as being the difference of the *I* values of two or more rectangles (as shown in Fig. 19.7) or the difference of the *I* values of two circles.

Consider the two examples shown in Fig. 19.8. In the first case, the *I* value for the I section can be determined by difference of *I* values for rectangular sections. In the second case, the *I* value for the hollow pipe is obtained by subtracting the *I* value for the inner circle from the *I* value for the outer circle. The calculations are given below.

For the I section shown in Fig. 19.8a:

$$I = \frac{BD^3}{12} - \frac{bd^3}{12} = \frac{150 \times 300^3}{12} - \frac{130 \times 260^3}{12} = (337.5 \times 10^6) - (190.4 \times 10^6)$$
$$= 147.1 \times 10^6 \text{ mm}^4$$

For the hollow pipe section shown in Fig. 19.8b:

$$I = \frac{\pi D^4}{64} - \frac{\pi d^4}{64} = \frac{\pi}{64}(D^4 - d^4) = \frac{\pi}{64}(300^4 - 270^4) = 137 \times 10^6 \text{ mm}^4$$

Calculation of second moment of area (*I*) for unsymmetrical sections

The bad news is that, for unsymmetrical sections, determination of the second moment of area (*I*) value is a whole lot trickier. In brief, the procedure for unsymmetrical sections is as follows:

1) Determine the position of the centroid of the section, using the approach outlined below. As you know from earlier in this chapter, the neutral axis always

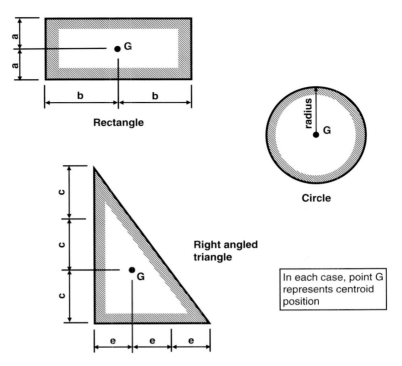

Fig. 19.9   Centroid positions for common shapes.

passes through the centroid of a section (assuming the section is made of the same material throughout). So, when you've determined the centroid position, you have also determined the level of the neutral axis.

2) Once you know the neutral axis position, use the ***parallel axis theorem*** (outlined below) to calculate the second moment of area (I) value.

Centroids and how to locate them

The centroid is the geometric centre of area of a body, shape or section. If a body is of uniform density, the centre of gravity will be at the centroid. If a structural element is homogeneous (i.e. of the same material throughout) and experiences pure bending, the neutral axis (i.e. axis of zero stress) will pass through the centroid. Therefore the location of the centroid of a cross section enables us to locate the level of the neutral axis (or neutral plane) relating to that cross section.

Figure 19.9 shows the centroid positions of some common shapes. As we can see, the centroids of rectangles and circles occur at the centre of area (i.e. the obvious point), whereas the centroid of a right-angled triangle occurs one-third of the way along each side from the right-angled corner – or two-thirds of the way along from a 'pointed corner'.

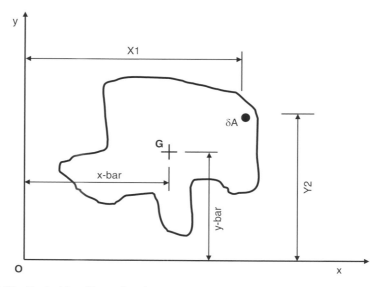

Fig. 19.10   Centroids of irregular shapes.

Centroids of irregular shapes

An irregular shape, and the location of its centroid, are indicated in Fig. 19.10, from which it can be shown that:

$$A\bar{x} = \sum (x.\delta A)$$

or:

$$\bar{x} = \sum \left(\frac{x.\delta A}{A} \right)$$

Similarly:

$$\bar{y} = \sum \left(\frac{y.\delta A}{A} \right)$$

where $\bar{x}$ and $\bar{y}$ are the distances from the y-axis and x-axis (respectively) to the centroid G. Note that the symbol Σ means 'sum of'. In other words, the dimension to the centroid of the total area from the appropriate axis or base line is equal to the sum of the area–distance products divided by total area.

Don't worry too much if you don't fully understand the mathematics above – it's the result and its application that are important.

Centroids of cross sections which can be broken down into regular shapes

Most cross sections encountered in civil engineering can be divided into constituent rectangles and triangles. The centroid positions in such cross sections may be calculated using the above formulas.

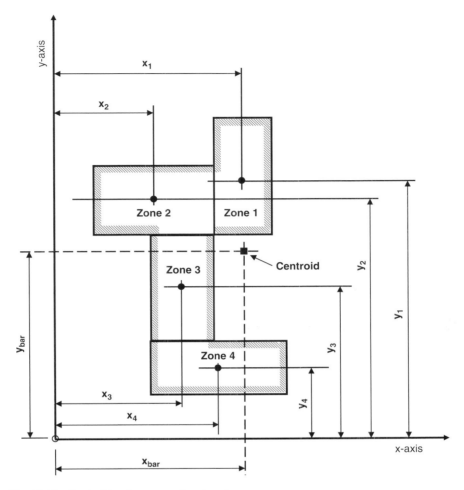

Fig. 19.11   Centroids of groups of rectangles.

Example

The beam shown in Fig. 19.11 can be divided into four rectangles as shown. The position of the section's centroid can be located from the following equations:

$$\overline{x} = \frac{A_1 x_1 + A_2 x_2 + A_3 x_3 + A_4 x_4}{A_1 + A_2 + A_3 + A_4}$$

$$\overline{y} = \frac{A_1 y_1 + A_2 y_2 + A_3 y_3 + A_4 y_4}{A_1 + A_2 + A_3 + A_4}$$

where:

A_1 = area of zone 1,

A_2 = area of zone 2, etc.

x_1 = distance from y-axis to centroid of zone 1

x_2 = distance from y-axis to centroid of zone 2, etc.

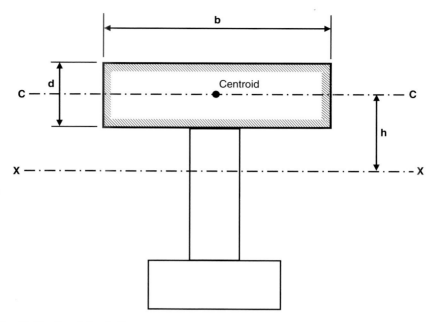

Fig. 19.12 Parallel axis theorem.

y_1 = distance from x-axis to centroid of zone 1
y_2 = distance from x-axis to centroid of zone 2, etc.

Parallel axis theorem

The ***parallel axis theorem*** can be used to calculate I values (i.e. second moment of area values) for sections that can be divided into individual rectangular parts. (For a rectangular section, $I = bd^3/12$.)

First, the neutral axis level (i.e. centroid position) has to be determined, in the manner previously discussed. Consider the rectangular element emphasised in Fig. 19.12, which forms part of a larger cross section. It can be shown that:

$$I_{XX} = I_{CC} + Ah^2$$

or, for a rectangle,

$$I_{XX} = (bd^3/12) + bdh^2$$

where:

I_{XX} = second moment of area of the rectangular element about the neutral axis of the composite section (i.e. about axis X–X)
I_{CC} = second moment of area of the rectangular element about the axis through its centroid (i.e. about axis C–C)
A = area of rectangular element
b = breadth of rectangular element

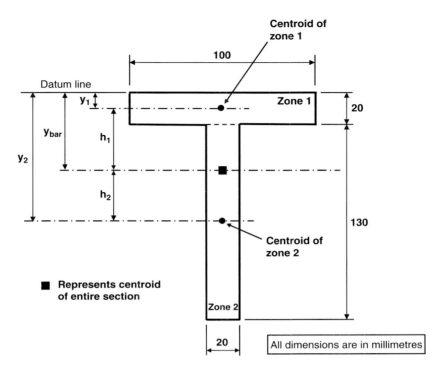

Fig. 19.13   Calculation of second moment of area for a T section (Example 19.3).

d = depth of rectangular element
h = distance between centroidal axis of the rectangular element and the centroidal axis of the composite section.

The total I_{xx} for the composite section is equal to the sum of the I_{xx} terms for the individual parts.

Example 19.3: Bending stresses in a T section

A beam with the cross section shown in Fig. 19.13 is simply supported and carries a maximum bending moment of 16.0 kN.m. Calculate:

- neutral axis position
- maximum tensile stress
- maximum compressive stress.

Solution

Use the top edge of the beam as a datum from which to calculate distances. I suggest a methodical way to approach this problem would be to do the calculations in tabular form – see Table 19.1.

In column 7: $42.4 = 52.4 - 10$
$32.6 = 85 - 52.4$

Table 19.1 Calculation of second moment of area for Example 19.3.

(1) Zone	(2) b (mm)	(3) d (mm)	(4) A (mm²)	(5) y (mm)	(6) Ay (mm³)	(7) h (mm)	(8) Ah² (mm⁴) (×10⁶)	(9) I = bd³/12 (mm⁴) (×10⁶)
1	100	20	2000	10	20 000	42.4	3.59	0.07
2	20	130	2600	85	221 000	32.6	2.76	3.66
Sum			4600		241 000		6.35	**3.73**

Split the cross section into two rectangles: let the cross-bar be zone 1 and the stem of the T be zone 2. The breadth (b) and depth (d) of each zone are given in columns 2 and 3 of Table 19.1. In each case, these are multiplied together to give the area of each zone, shown in column 4.

The y values are the vertical distances from the top of the section (the datum level) to the centroids of each zone. It can be seen from Fig. 19.13 that these values are 10 mm (half of 20 mm) for zone 1 and 85 mm (20 mm + half of 130 mm) for zone 2. These values are given in column 5 of Table 19.1.

The values given in column 6 are A (from column 4) multiplied by y (from column 5). From column 6 it can be seen that the sum of the Ay values is 241,000 mm³ and from column 4 the sum of the A values (i.e. the total area of the section) is 4600 mm². So the distance to the section's centroid from the top ($\bar{y}$) is calculated as follows:

$$\bar{y} = \frac{\Sigma(Ay)}{\Sigma A} = \frac{241,000 \text{ mm}^3}{4600 \text{ mm}^2} = 52.4 \text{ mm}$$

(This is the answer to part 1 of the question.)

Now that the position of the centroid of the section has been determined, the distances, h, from the section's centroidal axis to the centroids of the individual zones (depicted as h_1 and h_2 in Fig. 19.13) can be calculated. These figures are given in column 7 of Table 19.1.

Column 8, Ah^2, is A (from column 4) multiplied by h^2 (from column 7). Column 9 is the I value ($= bd^3/12$) for each rectangular zone.

When discussing the parallel axis theorem above, we saw that:

$$I_{XX} = Ah^2 + \frac{bd^3}{12}$$

So, I_{XX} is the sum of all the figures in column 8 and all the figures in column 9:

$$I_{XX} = (6.35 + 3.73) \times 10^6 = 10.08 \times 10^6 \text{ mm}^4$$

Now that we know the I value, we can calculate the bending stresses.
Earlier in this chapter we saw that:

$$\sigma = \frac{My}{I}$$

where:
    σ = bending stress
    M = bending moment
    y = distance from neutral axis to top or bottom of section
    I = second moment of area.

In this example:
    $M = 16.0\,\text{kN.m} = 16.0 \times 10^6\,\text{N.mm}$ (given in question)
    $I = 10.08 \times 10^6\,\text{mm}^4$ (calculated above)
    $y = 52.4\,\text{mm}$ (to top of section)
    $y = (150 - 52.4) = 97.6\,\text{mm}$ (to bottom of section)

As this beam is simply supported, the maximum tensile stress occurs in the bottom of the section and the maximum compressive stress occurs in the top. So:

Max tensile stress (bottom of section) =

$$\frac{My}{I} = \frac{16.0 \times 10^6\,\text{N.mm} \times 97.6\,\text{mm}}{10.08 \times 10^6\,\text{mm}^4} = 154.9\,\text{N}/\text{mm}^2$$

Max compressive stress (top of section) =

$$\frac{My}{I} = \frac{16.0 \times 10^6\,\text{N.mm} \times 52.4\,\text{mm}}{10.08 \times 10^6\,\text{mm}^4} = 83.2\,\text{N}/\text{mm}^2$$

Example 19.4: Bending stresses in a non-symmetrical I section

Determine the maximum bending moment that can be applied to the simply supported beam whose cross section is shown in Fig. 19.14 if:

- maximum tensile stress = $2.0\,\text{N/mm}^2$
- maximum compressive stress = $20\,\text{N/mm}^2$.

As in the previous example, we will use a table (Table 19.2) to calculate the neutral axis position and the second moment of area (I) value.

Again, we'll use the top edge of the beam as the datum. (Note: it doesn't matter what level you use as your datum, provided you are consistent throughout.)
From Table 19.2:

    In column 7: $372.7 = 387.7 - 15$
                       $57.7 = 387.7 - 330$
                       $262.3 = 650 - 387.7$

$$\bar{y} = \frac{\Sigma(Ay)}{\Sigma A} = \frac{17\,524\,000}{45\,200} = 387.7\,\text{mm from top (282.3\,mm from bottom)}$$

$$I_{xx} = (2604.33 + 543.07) \times 10^6 = 3147.4 \times 10^6\,\text{mm}^4$$

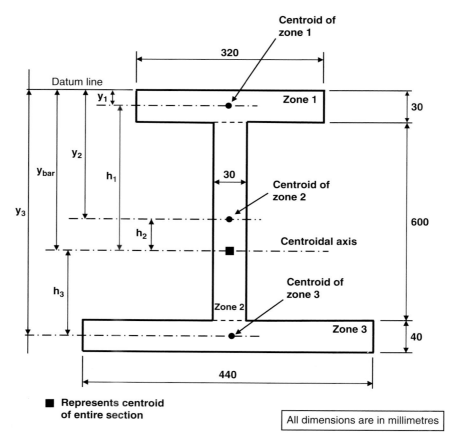

Fig. 19.14   Calculation of second moment of area for a non-symmetrical I section (Example 19.4).

Table 19.2   Calculation of second moment of area for Example 19.4.

(1)	(2)	(3)	(4)	(5)	(6)	(7)	(8)	(9)
Zone	b	d	A	y	Ay	h	Ah²	I = bd³/12
	(mm)	(mm)	(mm²)	(mm)	(mm³)	(mm)	(mm⁴) (×10⁶)	(mm⁴) (×10⁶)
1	320	30	9 600	15	144 000	372.7	1 333.50	0.72
2	30	600	18 000	330	5 940 000	57.7	59.93	540.00
3	440	40	17 600	650	11 440 000	262.3	1 210.90	2.35
Sum			45 200		17 524 000		2 604.33	**543.07**

Unlike the previous example, it is not bending stresses that we need to calculate this time. We need to determine the bending moments associated with particular values of tensile and compressive stress.

From the engineers' bending equation:

$$\frac{\sigma}{y} = \frac{M}{I}$$

Therefore, rearranging, we have:

$$M = \frac{\sigma y}{I}$$

Using the above equation we can calculate the moment that would cause the maximum (compressive) stress in the top of the beam and the moment that would cause the maximum (tensile) stress in the bottom:
In top of beam:

$$M = \frac{\sigma_{comp} \times I}{y_{top}} = \frac{20 \text{ N/mm}^2 \times 3147.4 \times 10^6 \text{ mm}^4}{387.7 \text{ mm}} = 162.4 \times 10^6 \text{ N.mm}$$
$$= 162.4 \text{ kN.m}$$

In bottom of beam:

$$M = \frac{\sigma_{tens} \times I}{y_{btm}} = \frac{2.0 \text{ N/mm}^2 \times 3147.4 \times 10^6 \text{ mm}^4}{282.3 \text{ mm}} = 22.3 \times 10^6 \text{ N.mm}$$
$$= 22.3 \text{ kN.m}$$

So the maximum bending moment that could be applied to the beam would be the lesser of the two figures calculated above, i.e. 22.3 kN.m.

What you should remember from this chapter

- A simply supported beam subjected to bending (in a sagging mode) will experience maximum tensile stress in the bottom, and maximum compressive stress in the top.
- The magnitude of the stress varies linearly between the top of the section and the bottom.
- The level at which there is no stress is called the neutral axis. For symmetrical sections of the same material throughout, the neutral axis occurs half way down the section.
- Before a given cross section can be analysed for stress, the second moment of area needs to be calculated. While this is relatively straightforward for symmetrical sections, it is more complicated for non-symmetrical sections, for which the parallel axis theorem must be used.

Tutorial questions

1) A timber beam of rectangular cross section 75 mm wide and 300 mm deep carries a 5 kN point load at the mid point of a simply supported span of 4 metres. Determine the maximum bending stress in the beam.
2) A steel beam with a symmetrical I-shaped cross section sustains a uniformly distributed load of 25 kN/m over a simply supported 3 metre span. The cross section dimensions (all in millimetres) are given in Fig. 19.15. Calculate:
   a) the maximum bending stress in the beam
   b) the radius of curvature of the beam, given $E = 205 \text{ kN/mm}^2$
   c) the bending stress at the top of the web in the beam at the location of maximum bending moment.

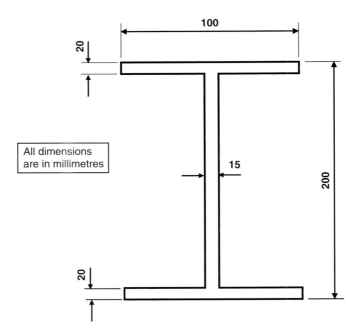

Fig. 19.15   Tutorial question 2.

3) A hollow tube of 50 mm external diameter and 44 mm internal diameter is subjected to a bending moment of 0.50 kN.m. Determine the maximum bending stress.

4) The reinforced concrete hollow rectangular section shown in Fig. 19.16 comprises the cross section of a 5 metre long beam, which sustains a 15 kN/m uniformly distributed load. Calculate:
   a) the maximum bending stress in the beam
   b) the radius of curvature of the beam given $E = 20$ kN/mm².

5) Figure 19.17 shows the cross-sectional geometry of a steel beam. The cross section of the beam is symmetrical about both the X–X and Y–Y axes. The beam spans 4 metres and supports a uniformly distributed load of 4 kN/m. Calculate:
   a) the second moment of area about the X–X axis (I_{xx})
   b) the maximum bending moment in the beam
   c) the maximum bending stress
   d) the strain corresponding to the stress calculated in c) if Young's modulus (E) for the steel beam is 205 kN/mm².

6) Figure 19.18 shows the geometry of a steel T section. A beam is constructed from this section and required to sustain a maximum bending moment of 75 kN.m. Calculate:
   a) the depth of the centroidal (X–X) axis from the top of the section
   b) the second moment of area about the X–X axis (I_{xx})
   c) the maximum bending stress in the beam when it is subjected to the maximum bending moment of 75 kN.m.

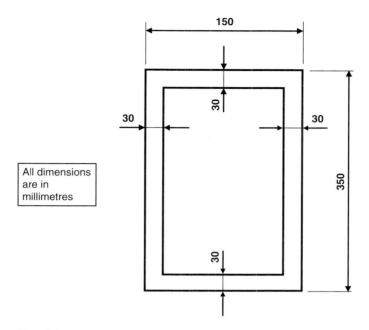

Fig. 19.16   Tutorial question 4.

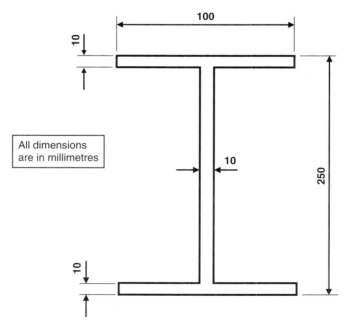

Fig. 19.17   Tutorial question 5.

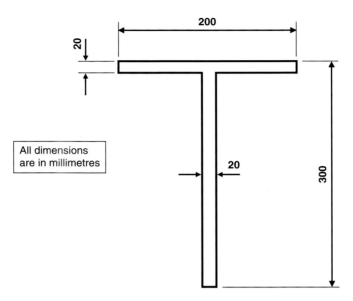

200

20

All dimensions
are in millimetres

20

300

Fig. 19.18 **Tutorial question 6.**

7) Figure 19.19 shows the cross section of a steel beam. The section is symmetrical about the Y–Y axis and the X–X axis passes through the centroid of the section and forms the neutral axis. Calculate:
   a) the depth of the centroidal (X–X) axis from the top of the section
   b) the second moment of area about the X–X axis (I_{xx})
   c) the maximum bending stress in the beam when it is subjected to a maximum bending moment of 50 kN.m.

Tutorial answers

1) 4.44 N/mm².
2) a) 74.6 N/mm²; b) 274.7 metres; c) 59.7 N/mm².
3) 101.8 N/mm².
4) a) 23.3 N/mm²; b) 150.5 metres.
5) a) 38.9×10^6 mm⁴; b) 8 kN.m; c) 25.7 N/mm²; d) 1.25×10^{-4}.
6) a) 97.5 mm; b) 89.2×10^6 mm⁴; c) 170.3 N/mm².
7) a) 85 mm; b) 16.35×10^6 mm⁴; c) 290.6 N/mm².

Suggestions for further work

Using a standard spreadsheet package, construct a spreadsheet that will calculate the following for any given T- or I-shaped cross section using the parallel axis theorem:

1) the depth from the top of the section to the neutral axis, in millimetres
2) the second moment of area of the section (I_{xx}) in mm⁴

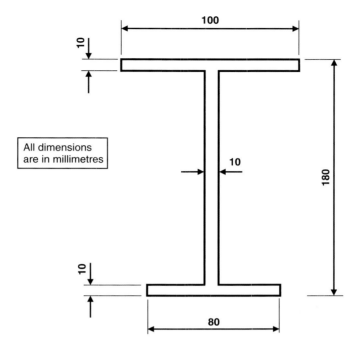

Fig. 19.19   Tutorial question 7.

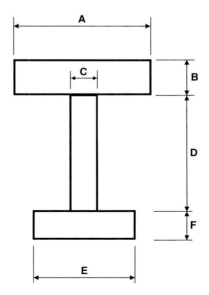

Fig. 19.20   Suggestions for further work.

3) for a given moment M (in kN.m), the bending stresses at the top of the section, the bottom of the section, and at the levels where the web meets the flange at the top and bottom.

Use your spreadsheet to check your answers to the tutorial questions above.

Extend your spreadsheet to calculate I_{xx} for symmetrical sections using the 'difference in area' method introduced earlier in this chapter, and show that the answers are the same as those using the parallel axis theorem.

(Hint: Label the dimensions A, B, C, D, E and F as shown in Figure 19.20. You can use the same spreadsheet for both T and I sections, because with a T section E and F will both be zero.)

20 Combined bending and axial stress

Introduction

In Chapter 18 we studied direct stresses. We found that the value of direct stress is constant across a cross section and is equal to the axial force (P) divided by the cross-sectional area (A). In Chapter 19 we investigated bending stresses. There we found that the value of bending stress is not constant across a cross section (in fact, it varies linearly) and that its maximum value is given by the bending moment (M) divided by its section modulus (z).

In this chapter we will see what happens when direct stresses and bending stresses are combined.

Combined stresses by formula

$$\text{Direct (axial) stress}(\sigma) = \frac{P}{A}\text{(from Chapter 18)}$$

$$\text{Maximum bending stress } (\sigma) = \frac{M}{z}\text{(from Chapter 19)}$$

These two equations can be combined, as shown below.

$$\text{Combined bending and axial stress} = \frac{P}{A} \pm \frac{M}{z}$$

It is not easy to see how this equation may be applied. To assist in this, look at Fig. 20.1. The two diagrams show the elevation of a column before and after an eccentric longitudinal load is applied. As you see, the left-hand side of the column (side A) is pushed down from its original position under the effects of the axial load, while the right-hand side of the column (side B) is pulled up. This suggests that side A is experiencing compression while side B is undergoing tension. However, it must be remembered that this bending effect (M/z) is in addition to the compression (P/A) caused by force P. So the bending effect *adds* to the compression on side A and *reduces* the compression on side B. If the force is sufficiently eccentric, side B can actually go into tension, i.e. if M/z is greater than P/A.

Basic Structures, Second Edition. Philip Garrison.
© 2011 John Wiley & Sons, Ltd. Published 2011 by John Wiley & Sons, Ltd.

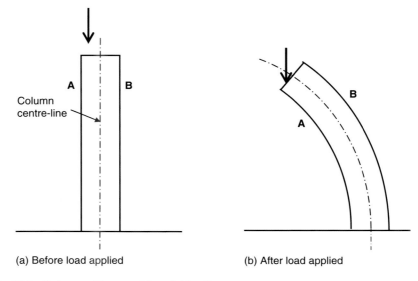

(a) Before load applied   (b) After load applied

Fig. 20.1   **Column with eccentric axial loading.**

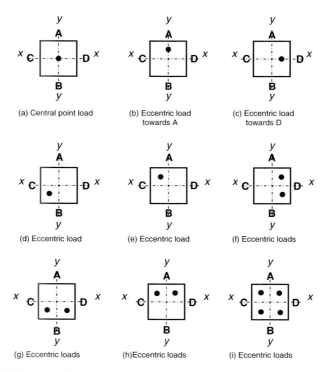

Fig. 20.2   **Eccentric loading on a column.**

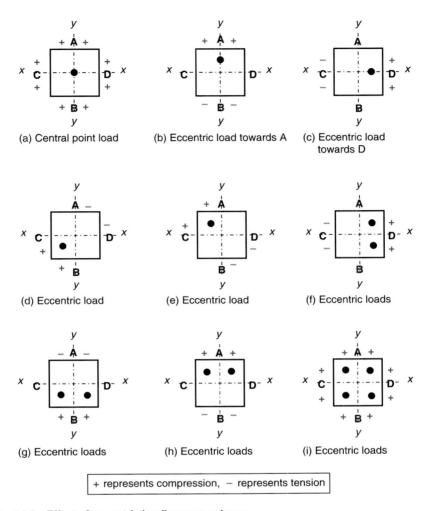

Fig. 20.3   **Effect of eccentric loading on a column.**

Each of the nine diagrams shown in Fig. 20.2 represents a plan view of a column which is square in cross section. The four sides are labelled A, B, C and D. In each diagram, the large black blob represents the position at which the longitudinal load is applied.

In each case, determine which side(s) of the column experience tension and which side(s) experience compression. To make things simpler, we will introduce a +/− sign convention as follows:

- If a side is pushed downwards under the applied load, it experiences compression (+).
- If a side is pulled upwards under the applied load, it experiences tension (−).

The answers, in the form of plus and minus signs, are shown in Fig. 20.3.

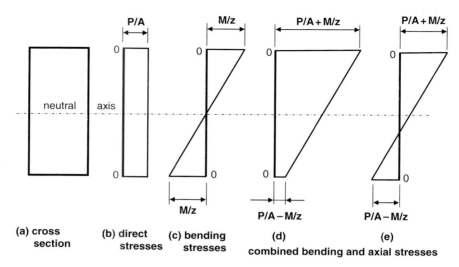

(a) cross section **(b) direct stresses** **(c) bending stresses** **(d)** **(e) combined bending and axial stresses**

Fig. 20.4 **Stress combinations.**

Keep the above exercise in mind as you progress through this chapter. It will help you to determine whether a plus or a minus sign is required at various points in your calculations.

Another way of looking at combined bending and axial stress

Consider the rectangular cross section shown in Fig. 20.4a. In Chapter 18 we learned that direct (or axial) stress has a value P/A which is constant across the cross section. This is illustrated in Fig. 20.4b. By contrast, we learned in Chapter 19 that the value of bending stress varies linearly across a cross section, with a maximum value of M/z. This was illustrated in Fig. 19.3 and is shown again, here, in Fig. 20.4c. If we combine the two graphs, the result depends on the relative values of P/A and M/z. If P/A is greater than M/z, the combined graph will appear as in Fig. 20.4d. But if P/A is less than M/z, the combination is shown in Fig. 20.4e. Note that this last case gives rise to tensile stresses when it is negative.

The formulas

I have said earlier in the book that I'm not a great fan of 'magic formulas' into which students can plug numbers and produce a (possibly incorrect) answer without a great deal of understanding of what they're doing. However, calculations for combined bending and axial stress situations are dependent on certain formulas – but you have to know when to use a plus sign and when to use a minus sign. And, as with any formula, you have to understand what the various terms mean.

Earlier in this chapter we encountered the following equation:

$$\text{Combined bending and axial stress} = \frac{P}{A} \pm \frac{M}{z}$$

Now, a force P acting at an eccentricity e from the centre line of a cross section will apply a moment of $(P \times e)$ at that centre line. So:

$$M = Pe$$

Also, in Chapter 19 we learned that $z = I/y$. From this we can generate two further equations for combined bending and axial stress, as follows:

$$\text{Combined bending and axial stress} = \frac{P}{A} \pm \frac{My}{I}$$

or:

$$\text{Combined bending and axial stress} = \frac{P}{A} \pm \frac{Pey}{I}$$

For a reminder of what all the symbols mean, see Fig. 20.5.

Example 20.1
A force of 200 kN acts vertically downwards on a column of cross-sectional dimensions 400 mm × 300 mm. The force acts at an eccentricity of 100 mm along the Y–Y axis from the centre of the section, as shown in Fig. 20.6a. Calculate the stress in the column at the following positions:

- along the 'top' face of the column (position A)
- at the point of application of the load (point K)
- at the centroid of the cross section (point L)
- at a point 50 mm 'below' the centre line (point M)
- along the 'bottom' face of the column (position B)

We know the following:

$$P = 200\,\text{kN (or } 200 \times 10^3\,\text{N)}$$

$$A = bd = (300\,\text{mm} \times 400\,\text{mm}) = 120{,}000\,\text{mm}^2$$

$$e = 100\,\text{mm}$$

$$M = Pe = (200 \times 10^3\,\text{N} \times 100\,\text{mm}) = 20 \times 10^6\,\text{N.mm}$$

$$I = \frac{bd^3}{12} = \frac{300 \times 400^3}{12} = 1.6 \times 10^9\,\text{mm}^4$$

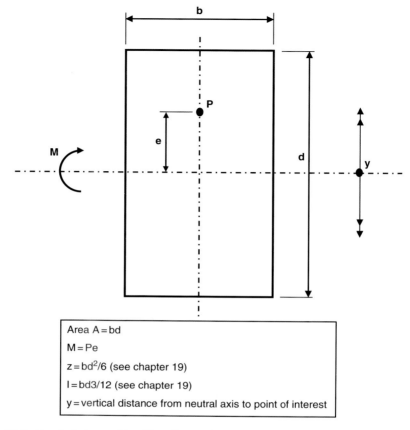

Area A = bd

M = Pe

z = bd²/6 (see chapter 19)

I = bd3/12 (see chapter 19)

y = vertical distance from neutral axis to point of interest

Fig. 20.5 Symbols in combined bending and axial stress equation.

y is the distance from the centroidal axis (X–X) to the position at which we're interested in calculating the stress. Its values for positions A, K, L, M and B are respectively 200, 100, 0, 50 and 200 mm.

Signs are also important. As the force P is pushing down on the upper part of the section, it will induce compression (+) for points A and K, zero for L, and tension (−) for points M and B.

$$\sigma = \frac{P}{A} \pm \frac{My}{I}$$

For point A: $\sigma_A = \dfrac{200 \times 10^3}{120,000} + \dfrac{20 \times 10^6 \times 200}{1.6 \times 10^9} = 1.67 + 2.5 = 4.17 \, \text{N/mm}^2$

For point K: $\sigma_K = \dfrac{200 \times 10^3}{120,000} + \dfrac{20 \times 10^6 \times 100}{1.6 \times 10^9} = 1.67 + 1.25 = 2.92 \, \text{N/mm}^2$

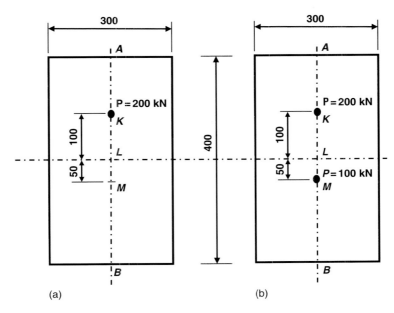

Fig. 20.6   Worked example 20.1.

For point L: $\sigma_L = \dfrac{200 \times 10^3}{120,000} + \dfrac{20 \times 10^6 \times 0}{1.6 \times 10^9} = 1.67 + 0 = 1.67 \, \text{N/mm}^2$

For point M: $\sigma_M = \dfrac{200 \times 10^3}{120,000} - \dfrac{20 \times 10^6 \times 50}{1.6 \times 10^9} = 1.67 - 0.625$

$$= 1.045 \, \text{N/mm}^2$$

For point B: $\sigma_B = \dfrac{200 \times 10^3}{120,000} - \dfrac{20 \times 10^6 \times 200}{1.6 \times 10^9} = 1.67 - 2.5 = -0.83 \, \text{N/mm}^2$

Now let's make the problem slightly harder. Let's suppose that, in addition to the 200 kN force shown above, a 100 kN force acts at point M, as illustrated in Fig. 20.6b. The overall moment about the X–X axis is now:

$$M = (200 \times 10^3 \, \text{N} \times 100 \, \text{mm}) - (100 \times 10^3 \, \text{N} \times 50 \, \text{mm}) = 15 \times 10^6 \, \text{N.mm}$$

The total force, P, is now:

$$(200 \, \text{kN} + 100 \, \text{kN}) = 300 \, \text{kN} \ (\text{or} \ 300 \times 10^3 \, \text{N})$$

So, the first term of the equation is now:

$$\frac{P}{A} = \frac{300 \times 10^3}{120,000} = 2.5 \, \text{N/mm}^2$$

Table 20.1   Stresses derived from Example 20.1.

Point	Description of point	stress (in N/mm²) for 200 kN load only	stress (in N/mm²) for 200 kN load + 100 kN load
A	'Top' face of column	+4.17	+4.375
K	100 mm above centre	+2.92	+3.438
L	At centre of column section	+1.67	+2.5
M	50 mm below centre	+1.045	+2.453
B	'Bottom' face of column	−0.83	+0.625

The other quantities remain the same. So now the stresses are as follows:

$$\text{For point A: } \sigma_A = 2.5 + \frac{15 \times 10^6 \times 200}{1.6 \times 10^9} = 2.5 + 1.875 = 4.375 \, \text{N/mm}^2$$

$$\text{For point K: } \sigma_K = 2.5 + \frac{15 \times 10^6 \times 100}{1.6 \times 10^9} = 2.5 + 0.938 = 3.438 \, \text{N/mm}^2$$

$$\text{For point L: } \sigma_L = 2.5 + \frac{15 \times 10^6 \times 0}{1.6 \times 10^9} = 2.5 + 0 = 2.5 \, \text{N/mm}^2$$

$$\text{For point M: } \sigma_M = 2.5 - \frac{15 \times 10^6 \times 50}{1.6 \times 10^9} = 2.5 - 0.047 = 2.45 \, \text{N/mm}^2$$

$$\text{For point B: } \sigma_B = 2.5 - \frac{15 \times 10^6 \times 200}{1.6 \times 10^9} = 2.5 - 1.875 = +0.625 \, \text{N/mm}^2$$

These stresses, for each of the two cases considered, are tabulated in Table 20.1.

Beware the difference between e and y

Many students are puzzled as to the distinction between e and y. This distinction is crucial to the understanding of problems involving combined bending and axial stress, and is as follows:

- e represents the eccentricity of the load(s) – that is, the distance from the point of action of the load to the relevant centroidal axis (axis X–X in the above case). In the above example, $e = 100$ mm for the 200 kN load and 50 mm for the 100 kN load.
- y represents the distance from the centroidal axis to the point at which we wish to know the stress.

Maximum and minimum values of stress

Examine the figures in Table 20.1. You will see that in each case the maximum stress occurs in the 'top' face (position A) and the minimum stress occurs in the 'bottom' face (position B).

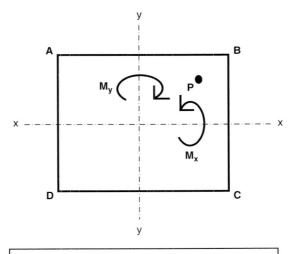

Moment M_x is clockwise when viewed from side BC
Moment M_y is clockwise when viewed from side DC

Fig. 20.7   Stresses caused by rotation about both axes – general case.

The values of maximum and minimum stresses are particularly important to engineers, as we design a column (or other structural element) to sustain the worst stress to which it is likely to be subjected. This 'worst' stress is usually the maximum value, but the minimum value is of interest too, especially if it is negative (as it was in the case of point B in Example 20.1 above). A negative value of stress suggests that tension is being experienced, and in many situations we need to avoid tensile stresses. More of that later.

Combined stresses in two dimensions

So far we have considered stresses in one dimension only. (For example, in Fig. 20.6, points A, K, L, M and B all lie on the same line.) This is fine for situations where the loads happen to act directly on centroidal axes, but what happens if they don't?

Examine Fig. 20.7, which shows a column cross section on which a load P is acting. P acts at a point which is eccentric from the column's centroid in both directions – in other words, the point is not on either the X–X or Y–Y axes. The four corners of the column are labelled A, B, C and D.

The eccentric load P will induce a moment about each of the axes X–X and Y–Y. We will call these moments M_x and M_y respectively.

$$z_x = bd^2/6 \text{ and } z_y = db^2/6$$

(The section modulus, z, was introduced in Chapter 19.)

The stresses at the four corners (A, B, C and D) of the column can be calculated from the following equations:

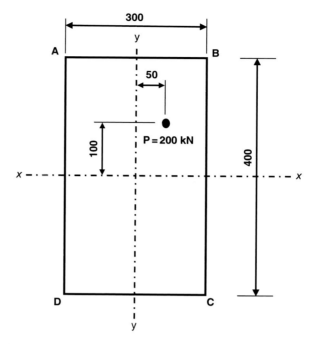

Fig. 20.8   **Worked example 20.2**

$$\sigma_A = \frac{P}{A} + \frac{M_x}{z_x} - \frac{M_y}{z_y}$$

$$\sigma_B = \frac{P}{A} + \frac{M_x}{z_x} + \frac{M_y}{z_y}$$

$$\sigma_C = \frac{P}{A} - \frac{M_x}{z_x} + \frac{M_y}{z_y}$$

$$\sigma_D = \frac{P}{A} - \frac{M_x}{z_x} - \frac{M_y}{z_y}$$

Example 20.2

As in Example 20.1 above, a column of cross-sectional dimensions 400 mm × 300 mm experiences a load of 200 kN. This time though, the load is applied eccentrically to both axes, as shown in Fig. 20.8. Calculate the stress at each of the four corners of the column (A, B, C and D).

$$P = 200\,\text{kN (or } 200 \times 10^3\,\text{N)}$$

$$A = (300\,\text{mm} \times 400\,\text{mm}) = 120{,}000\,\text{mm}^2$$

$$M_x = +(200 \times 10^3 \, \text{N} \times 100 \, \text{mm}) = 20 \times 10^6 \, \text{N.mm}$$

$$M_y = +(200 \times 10^3 \, \text{N} \times 50 \, \text{mm}) = 10 \times 10^6 \, \text{N.mm}$$

$$z_x = bd^2/6 = 300 \times 400^2/6 = 8.0 \times 10^6 \, \text{mm}^3$$

$$z_y = db^2/6 = 400 \times 300^2/6 = 6.0 \times 10^6 \, \text{mm}^3$$

Note that M_x and M_y are both positive because they both act in the same direction as the general case shown in Fig. 20.7.

$$\sigma = \frac{P}{A} \pm \frac{M_x}{z_x} \pm \frac{M_y}{z_y}$$

$$\sigma = \frac{200 \times 10^3}{120,000} \pm \frac{20 \times 10^6}{8 \times 10^6} \pm \frac{10 \times 10^6}{6 \times 10^6} \, \text{N} / \text{mm}^2$$

$$\sigma = 1.67 \pm 2.5 \pm 1.67 \, \text{N/mm}^2$$

So the stresses at the four corners are:

$$\sigma_A = 1.67 + 2.5 - 1.67 = +2.5 \, \text{N/mm}^2$$

$$\sigma_B = 1.67 + 2.5 + 1.67 = +5.84 \, \text{N/mm}^2$$

$$\sigma_C = 1.67 - 2.5 + 1.67 = +0.84 \, \text{N/mm}^2$$

$$\sigma_D = 1.67 - 2.5 - 1.67 = -2.5 \, \text{N/mm}^2$$

Note the negative value of stress at point D – it indicates that tensile stress is being experienced there.

Pressure on foundations

The principles outlined above regarding eccentric loads on columns are equally applicable to eccentric loads on foundations. Columns in buildings have to be supported at their base by a foundation, whose function is to safely transmit all the loads from a structure into the ground (see Chapter 1). A concrete pad (or isolated) footing is often used, as illustrated in Chapter 3.

In the design of pad foundations it is important to ensure that the permissible ground bearing pressure (that is, the maximum pressure that the ground can sustain) is not exceeded. It is therefore important to be able to calculate the actual pressure at any point in the foundation. In practice, the maximum or minimum pressures occur at one of the four corners, so it is sufficient to calculate the actual pressure at each of the corners.

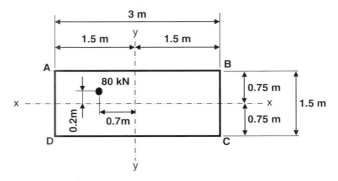

Fig. 20.9   Example of eccentric foundation loading.

Figure 20.7, which we referred to earlier when we were considering stresses in columns, is equally applicable to the general case for pressure on foundations. It is a plan view of a rectangular concrete pad foundation whose four corners are labelled A, B, C and D.

The two centroidal axes are labelled X–X and Y–Y. An eccentric load P acts at a position that causes a clockwise moment (as viewed from side BC) about axis X–X and a clockwise moment (as viewed from side DC) about axis Y–Y. The pressures at the corners A, B, C and D are given by the four equations discussed earlier.

(Note: Although we used N and mm units when calculating stresses in columns, the larger forces and dimensions in foundations suggest that kN and metres are more suitable units when calculating pressures in foundations.)

Example 20.3
Calculate the pressure at each corner of the foundation shown in Fig. 20.9. The 80 kN load will cause a clockwise rotation about the x-axis (as viewed from side BC) which is the same as that assumed in the general case of Fig. 20.8. Hence the positive sign in the M_x calculation below.

The 80 kN load will cause an anticlockwise rotation about the y-axis (as viewed from side DC), which is opposite in direction from the clockwise rotation assumed in the general case of Fig. 20.8. Hence the negative sign in the M_y calculation below.

$$P = 80\,kN$$

$$M_x = +(80\,kN \times 0.2\,m) = +16\,kN.m$$

$$M_y = -(80\,kN \times 0.7\,m) = -56\,kN.m$$

$$A = (3.0 \times 1.5) = 4.5\,m^2$$

$$z_x = \frac{bd^2}{6} = \frac{3 \times 1.5^2}{6} = 1.125\,m^3$$

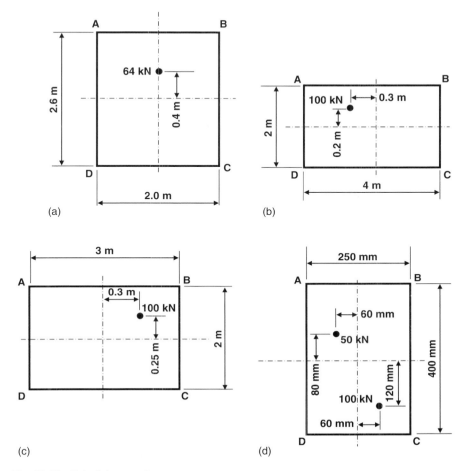

Fig. 20.10   Tutorial examples.

$$z_x = \frac{db^2}{6} = \frac{1.5 \times 3^2}{6} = -2.25\,\text{m}^3$$

$$\sigma = \frac{P}{A} \pm \frac{M_x}{z_x} \pm \frac{M_y}{z_y}$$

$$\sigma = \frac{80}{4.5} \pm \frac{16}{1.125} \pm \frac{-56}{2.25}$$

$$\sigma_A = 17.78 \pm 14.22 \pm (-24.89)$$

$$\sigma_A = 17.78 + 14.22 - (-24.89) = +56.89\,\text{kN/m}^2$$

$$\sigma_B = 17.78 + 14.22 + (-24.89) = +7.11 \, \text{kN/m}^2$$

$$\sigma_C = 17.78 - 14.22 + (-24.89) = -21.33 \, \text{kN/m}^2$$

$$\sigma_D = 17.78 - 14.22 - (-24.89) = +28.45 \, \text{kN/m}^2$$

As the pressure at corners C is negative, this suggests that tension occurs at this point. In other words, the foundation would tend to lift off the ground at point C, which is obviously not desirable in practice!

What you should remember from this chapter

This chapter explains how to combine axial and bending stresses in a column or a foundation. The calculation procedure has been outlined to obtain the overall stress (or pressure) at any point in a column (or foundation). Watch out for the signs (+ or −) and be aware that a negative stress indicates that tension is occurring at the point concerned.

Tutorial questions

Calculate the stresses at each of the four corners (A, B, C and D) of the four examples illustrated in Fig. 20.10. In each case, identify the points (if any) at which tension occurs. (Note: In each case, the loads act 'into the paper'.)

Tutorial answers

Values given are at points A, B, C and D respectively.

a)   +23.67, +23.67, +0.95, +0.95 kN/m².
b)   +25.63, +14.37, −0.63, +10.63 kN/m².
c)   +19.17, +39.17, +14.17, −5.83 kN/m².
d)   −419, +1019, +3419, +1981 kN/m².

21 Structural materials: concrete, steel, timber and masonry

Introduction

This book is primarily concerned with the basics of structural analysis. Up till now we haven't paid much attention to the material that a beam, column or slab might be made of. There are, of course, many materials available for us to use, but in this chapter we will confine our discussion to the four main structural materials, namely concrete, steel, timber and masonry.

Both architects and structural engineers need to decide at an early stage what material (or combination of materials) they are going to use in a particular project. But it's difficult to make such a decision if you don't know anything about the various materials. The purpose of this chapter is to discuss the different materials available to the construction professional.

Which is the best material?

A natural question at this stage is: which is the best structural material? Well, it depends on what you mean by 'best'. Does 'best' mean strongest, stiffest, cheapest, most readily available or most attractive? Or all of these? Or maybe none of these?

A moment's consideration would lead us to conclude that there is no one building material that is the best in all respects. If there were, then every building structure in the entire world would be built out of that one material. Clearly this isn't the case. If we look at the world around us, we see buildings made of brickwork or stonework, timber buildings, and buildings with frames of steel or reinforced concrete. In certain parts of the world we see buildings constructed of ice, mud or bamboo. It is apparent that there are many different materials that can be used in building, each of which has its advantages and disadvantages.

The kettle analogy

If you look at your everyday surroundings you will notice that particular objects tend to be made of certain materials. This is because these materials are particularly suitable for given applications. For example, car tyres are made of rubber, windows are made of glass, pens are usually plastic.

Basic Structures, Second Edition. Philip Garrison.
© 2011 John Wiley & Sons, Ltd. Published 2011 by John Wiley & Sons, Ltd.

We also know that certain materials are patently unsuitable for certain applications. For example:

- contact lenses are never made from steel
- aircraft fuselages are never constructed from brickwork
- computers are never made out of concrete
- radiators are never made from plastic (although perhaps they could be).

Consider a *kettle* as an example. If you review the desired properties of a kettle, you might come up with some or all of the following:

- Strength: the kettle must be strong enough to contain water and to resist the pressure of steam building up inside it. It must also be strong enough not to break if dropped onto a hard floor surface.
- Thermal properties: the kettle must be able to resist the temperature of boiling water and must not break, melt or otherwise deform at such temperatures. It must also be able to cope with sudden changes of temperature, for example if cold water is poured into a recently boiled kettle.
- Rigidity: the kettle must not deform under water or steam pressure.
- Disposability: what will happen to the kettle at the end of its life?
- Availability of materials: the materials must be readily available in the quantities required for mass production of kettles.
- Manufacturing costs: the manufacturing process must be streamlined so that kettles are produced as cheaply as possible.
- Durability: the kettle should not readily rot, corrode or otherwise degrade in use.
- Waterproofness: the kettle shouldn't leak.
- Attractiveness: the kettle should be sufficiently good looking that people would want to buy it.

A manufacturer of kettles has to find a material that has all the above properties. Until the late 1970s, all kettles were made of steel; then plastics were developed that could cope with high temperatures without deforming. Nowadays most kettles are made of plastic because there are plastics available that meet the above requirements and are cheaper than steel. Let's consider the consequences of making kettles out of other materials.

- A timber kettle is possibly more expensive to manufacture. It would be difficult to achieve a waterproof seal and the timber would rot quickly in such a damp, steamy environment unless preservatives were used – which may be poisonous!
- It would be difficult (and therefore uneconomic) to create a concrete kettle to the required dimensions; otherwise it would be too heavy. Also, the surface of the concrete might tend to flake off or dissolve into the water being boiled.
- A masonry kettle would be impractical for the same reasons as a concrete one, with formation of waterproof joints being an additional problem.

So what was the purpose of this diversion into the preferred properties of a kettle? Well, some of the properties listed above, desirable in the manufacture of kettles, are also important properties of the materials to be used in structures. Let's examine some of these desirable properties in more detail.

Factors to be considered in material selection

Availability

Construction materials are used in large quantities and therefore need to be readily available. Stone and clay are extracted in most parts of the UK, hence masonry (stonework, brickwork and blockwork) is widely used in domestic construction. (For example, until the 1960s every building in the Scottish city of Aberdeen was built out of granite, which was readily available locally from one massive quarry.) In some parts of the world, other locally available materials are excellent for construction. Also, the local labour force is likely to be familiar with the use of locally available materials.

Strength

Materials need to be strong enough (in tension and/or compression) for their intended purpose. Clearly, some materials are stronger than others. Selection of too weak a material for a particular application is an obvious mistake, but selection of a needlessly strong material is also undesirable.

Stiffness

Stiffness, or rigidity, is not to be confused with strength: some strong materials are not stiff (e.g. rope) and some stiff materials are not particularly strong (e.g. glass). The stiffer a material, the less it will deflect. The stiffness of a material is proportional to its Young's modulus value. (For the derivation of Young's modulus, see Chapter 18.) Typical Young's modulus values for the materials being considered in this chapter are:

- Steel: $210\,kN/mm^2$
- Aluminium: $71\,kN/mm^2$
- Concrete: $14\,kN/mm^2$
- Timber: $5–10\,kN/mm^2$

It can be seen from the above that steel is by far the stiffest of the common structural materials – for a given cross section steel is three times as stiff as aluminium, 15 times as stiff as concrete and over 20 times as stiff as timber. But remember, this is for a constant cross section, so these relative stiffnesses will vary according to the cross section used.

We saw in Chapter 1 that deflection needs to be controlled, but it is less critical in some applications than others. A super-stiff material, therefore, is not always required or even desirable.

Speed of erection

Some building types can be erected more quickly than others. For example, a steel-framed structure can be completed far more quickly than a masonry one. But speed of construction is not always critical and there may well be a trade-off between speed and cost. Being told that a building could be built twice as fast for twice the cost greatly concentrates the mind!

Cost/economics

A complex issue. Architects and engineers are always looking to minimise cost. There is an old saying that an engineer can do for a penny what anyone can do for two pence. We have to consider the cost of the raw materials, the cost of conversion of the material into its usable form, transportation costs and associated labour costs.

Ability to accommodate movement

All buildings tend to move. Some materials can accommodate this better than others. For example, brickwork can cope with movement more readily than a steel-framed structure can.

Durability

Some materials rot, decompose, corrode or spall, etc. over time. Some materials do this more readily than others; in other words, some materials are less durable than others. Maintenance costs and programmes need to be taken into account. For example, it is well-known that the Forth Rail Bridge in Scotland is repainted on a three- to five-year cycle to control corrosion of the steel structure.

Disposal

Nothing lasts for ever. How is the building going to be disposed of at the end of its life? Can the material be reused or converted into some other usable form? What are the costs associated with this?

Fire protection

There is an unfortunate possibility that any structure may catch fire. Some materials have better fire resistance properties than others.

Size and nature of the site

The location of the site may influence the choice of materials. Traffic congestion problems, local by-laws and physical obstructions may limit the size of deliveries to the site and the times of day that deliveries can take place.

We will now discuss each of the main structural materials individually. As you will see, each material has its advantages and disadvantages.

Concrete

Concrete is manufactured by mixing four ingredients – cement, fine aggregate (sand), coarse aggregate (gravel or crushed rock) and water – in predetermined proportions in a controlled manner to form a grey fluid resembling porridge. This wet concrete is transported to the place where it is needed and poured into 'moulds' of the required shape and size. These moulds, known by the terms *formwork* or *shuttering*, are usually made of timber or steel. Chemical reactions take place within the concrete, which lead to its setting, hardening and gaining in strength over a period of weeks.

The production of concrete needs to be carefully controlled. Firstly, its naturally occurring constituent materials are variable in quality. Secondly, wet concrete is susceptible to high or low temperatures and needs to be placed as quickly as possible before it 'goes off' (that is, sets). Thirdly, careless treatment of wet concrete – for example, allowing it to drop from a great height or to bounce off formwork – can lead to segregation of its constituents which can affect the integrity of the finished concrete.

Concrete is strong in compression (typically 30–40 N/mm^2) but weak in tension (3–8 N/mm^2). As we saw in Chapter 3, any structural element in bending – for example, a beam or a slab – experiences tension, therefore, if an element is made of concrete, it needs to be reinforced with steel bars. Concrete with steel bars in it is known as *reinforced concrete*. In practice, all concrete seen in structures is reinforced concrete.

Reinforced concrete has a number of advantages:

- It has high strength when reinforced.
- It is mouldable into any desired shape.
- Because it is mouldable, it can be formed into structurally continuous elements.
- It is durable: it does not corrode or rot.
- It has good fire resistance properties.
- It also has good thermal and noise insulation properties.
- It is relatively cheap to produce – although its placement on site is quite labour-intensive, which increases the cost.
- It can be used compositely (that is, two materials acting together) with structural steel.
- It is widely used in foundations, columns, beams, slabs, bridges, roads and railway sleepers.
- It is suitable for short-span low- and high-rise building frames.
- Prestressed concrete – concrete through which highly tensioned rods or cables have been placed – is stronger than reinforced concrete and therefore longer and more slender members can be used. Prestressed concrete is therefore suitable for long spans and rigid frames. You'll read more about prestressing in Chapter 25.
- Concrete elements (beams, columns, etc.) can be made in factories and then, when hardened, transported to a construction site and erected into position. Such elements are termed *precast* concrete elements. The more usual concrete construction, where wet concrete is poured into formwork on site, is called *in-situ* construction.

However, the following disadvantages of reinforced concrete also need to be considered:

- It is heavy, both physically and aesthetically.
- As indicated above, construction using reinforced concrete needs to be carefully controlled and is labour intensive. It is 'messy', requiring formwork, reinforcement and the placing and compaction of the concrete.

- Once poured, it takes several weeks for the concrete to achieve the required strength. This delays consequent construction activities (unless the concrete is precast).
- Although it doesn't rot or corrode, concrete can suffer certain ills, including spalling, cracking (leading to possible corrosion of reinforcement) and carbonation (a chemical reaction with the atmosphere that causes deterioration).

Masonry

Traditionally the term masonry refers to the material crafted by a mason – namely, stone. In modern times, the term more usually applies to brickwork or blockwork.

Bricks and blocks come in small, cuboidal units which can be lifted by hand. They are laid in rows by a bricklayer to form walls or columns. Mortar is used to 'glue' the individual units together and to fill the gaps or any irregularities between units.

The advantages of masonry are as follows:

- It has high compressive strength, making it ideal for walls, columns and arches, all of which are in pure compression.
- It is durable – no finish is required.
- It is made from raw materials readily available in the UK at low cost.
- No complicated plant is required.
- It has an attractive appearance.
- There is design flexibility – bricks or blocks can be combined to form complex shapes.
- Masonry has good fire resistance properties and good thermal/acoustic properties.

The disadvantages of masonry are as follows:

- It has very low tensile strength, which means that it that cannot be used for elements that bend, for example beams or slabs.
- Compared with timber (the other material used for low-rise domestic construction), masonry is heavy, so larger foundations are required and transport costs are higher.
- Frost and chemical attack can cause spalling in brickwork.
- Efflorescence – chalky and unsightly (but harmless) deposits – can occur on brickwork following a cycle of wetting and drying.

Because of its durability masonry buildings have excellent potential for change of use. Figure 21.1 shows a traditional stone church in the Dutch city of Maastricht which is now enjoying a new life as a bookshop.

Timber

Timber is the only structural material which is used in its naturally occurring form. The length and cross section of a timber beam are limited by the height and girth of the tree from which it is obtained.

Fig. 21.1   A church becomes a bookshop.

Longer timber beams, and larger cross sections, can be obtained by slicing the timber into thin strips and gluing these strips together both along their lengths and their ends, but it is an expensive process rarely used in the UK. This is known as glued laminated (or 'glulam') timber.

Timber comes in two types:

- hardwoods, obtained from deciduous (leaf-shedding) trees
- softwoods, obtained from coniferous (evergreen) trees.

Softwoods are generally used for structural purposes.

Timber is one of the oldest building materials and has the following structural advantages:

- It is light, with a high strength/weight ratio.
- It is easy to cut and shape.
- Despite what you might expect, it performs well in fire.
- It has good chemical durability.
- It has a pleasing appearance.
- It is relatively cheap.
- Although it has low stiffness, it is relatively stiff in relation to its own (light) weight.
- It is suitable for lightly or moderately loaded low-rise building frames and for shed and rigid frames.

But timber has the following disadvantages:

- Its low strength means that spans are limited, as is the height of timber buildings.
- It is difficult to form joints in certain circumstances.
- As mentioned above, the size of a piece of timber is limited by the size of the tree from which it comes.
- Timber is susceptible to rot and decay unless properly maintained.
- Its properties vary according to species of tree.

Steelwork

Structural steelwork is manufactured in standard sections. It has the following advantages:

- Its strength is high in both tension and compression (but compression in steelwork can be a problem – see below).
- Steel has a high strength/weight ratio.
- Because steel sections are produced in a factory under carefully controlled conditions, high quality control can be achieved.
- Steel's appearance can be elegant, with slender elements, smooth surfaces, straight and sharp edges.
- Prefabrication is possible.
- Steel has high stiffness.
- Steel is economic in material: a small amount carries a relatively large load.
- Steel is suitable for low/high-rise buildings and roof structures of all spans.

Steelwork does, however, have the following disadvantages:

- It is heavy: cranes are required to lift steelwork.
- It is a high-cost material.
- It has a durability problem: it corrodes if not protected and maintained.
- It has poor fire resistance; therefore steelwork needs to be protected by other materials.
- Because of the slender sections used in steelwork, it is prone to buckling in compression. This is an important criterion in the design of steelwork.

The SAGE in Gateshead in northern England, shown in Figure 21.2, is a performing arts centre which comprises three separate concert hall buildings enclosed by a steel and glass shell which curves in three dimensions, necessitating some complex steelwork fabrication. Its detractors have likened the building to a giant slug.

Aluminium

Aluminium is rarely used as a structural material except in very small structures (e.g. greenhouses). Its main properties are as follows:

- Its strength is about the same as mild steel.
- It is stiffer than concrete or timber.

Fig. 21.2   **SAGE, Gateshead.**

- It is less stiff than steel, but also lighter.
- It has a high strength/weight ratio.
- But: aluminium is expensive.

So how do I decide what materials to use in a given building?

The following discussion relates to construction in the UK, though some of it may apply elsewhere.

Framed or unframed structure?

The first decision to be made is whether the structure will be framed or unframed. In a framed structure, a framework or 'skeleton' of beams and columns is used to carry the structural loads down the building to the foundations. The framework is usually of steel or reinforced concrete, but in very small (usually single-storey) structures it may be of timber or aluminium. The finished building will usually also have external and internal walls, but these are non-structural and support no loads other than their own weight.

In a non-framed structure, the walls are load-bearing. These load-bearing walls are usually masonry, but may be reinforced concrete.

Example 21.1

Consider the following scenario.

Depending on your specialism, you run either an architectural practice or a firm of consulting engineers. One of your clients, a property development company, proposes to construct an office development on a specific site. Dimensions of the planned building have yet to be finalised, but it is known that the building will be two-storey, of approximate plan dimensions 60 m × 20 m. When complete, the building will be rented out to either one company or, with appropriate subdivisions, to a number of small tenant companies.

At the first meeting of the project team, your client asks your advice on whether a framed structure would be appropriate. Write your reply, giving full reasons for your choice.

Having thought about this, your answer would probably be that a framed structure is the appropriate option, for the following reasons:

- It is clear that the use of the building is not rigidly defined. It is an office building, but it may be occupied by a number of companies, and the tenant companies may grow (thus requiring more space) or shrink (requiring less space). Companies may come and go over time. Accordingly, the available space should be as flexible as possible to accommodate the changing needs of the tenants. It is best not to have such flexibility inhibited by the presence of internal load-bearing walls.
- The absence of load-bearing walls means there will be more floor space. Although this increase in floor space will be relatively small, it will be good news for your property developer client, who will be anxious to squeeze as many lettable square metres as possible out of the building.
- If there are no load-bearing walls – which would be made of concrete or masonry and so would be relatively heavy – the building as a whole will be lighter. This relative lightness would mean that the loads on the foundations would be less, which in turn means that the foundations could be less substantial and thus cheaper. Your client would be delighted at any saving in money that you could offer.
- Framed structures of steel or concrete can be erected much faster than load-bearing masonry structures. This will again please your client, who will want to see the structure completed (and thus providing rental income) as soon as possible – preferably yesterday.

However, as with most projects in 'the real world', things do not run smoothly and there is a twist in the tale:

At the second meeting of the project team, your client shares his belief that a forthcoming recession will cause a drastic decrease in the demand for office accommodation. He does, however, foresee a growing demand for quality hotel accommodation and has therefore replaced the office project with a hotel project on the same site, which, when complete, will be sold to the Dream Easy Inn hotel chain for use as a bedroom block. Due to planning constraints, the height and overall dimensions of the building will remain as before.

Your client asks whether this change of use would change your earlier advice on the building's structure. What is your reply? Give reasons.

Now the scheme has changed totally. Although the final building will be the same shape and size as before, its use is now completely different. The needs of a hotel chain (and the guests who pay to stay there) are vastly different from the demands of a company renting office space (and those of the office workers it employs). So the architect and engineer need to think again.

In this case, you may well decide that a framed structure is not appropriate, for the following reasons:

- Guests in a hotel room want a good night's sleep. It is therefore important that the hotel room be at the right temperature and quiet – no guest wants to be disturbed by noise from the room next door or from outside. High levels of thermal and sound insulation are therefore important. It makes sense to use load-bearing blockwork which, correctly specified, would provide an appropriate level of thermal and sound insulation as well as forming part of the building's structure.
- Unlike the office scenario, no flexibility is required of a hotel bedroom block. It is unlikely that the hotel owner would need to change the size of individual hotel rooms or the location of their walls in the future.
- Once again, you should consider your client's needs. As he will be selling the building on to a hotel chain on completion, his main concern is that the finished building will be an attractive purchase for such an operator. Your client is not concerned about the building's future income potential.
- It should be noted that this building is low-rise (only two storeys). The decision might be different with a high-rise building, where the efficiency of a structural framework would override other considerations.

We can extrapolate the lessons we've learned from this specific example to more general cases, as follows.

- Features of framed structures:
- flexibility: can accommodate change of use
- small saving in floor space
- lighter, giving smaller (and hence cheaper) foundations
- faster speed of erection

Features of non-framed structures:

- inherent thermal and sound insulation properties in masonry, so useful for hotels or apartment buildings where insulation is important
- no flexibility in the use of the building – but this may not be required anyway

The following is a list of the materials used for particular structural elements.

Walls

- masonry (unframed structures).
- masonry, timber stud, aluminium frame (framed structures)

Floors

- timber joists supporting floorboards (domestic: low loads, small spans)
- in-situ reinforced concrete (general industrial/commercial)
- precast concrete (suitable for regular, repetitive floor layouts)
- composite: in situ concrete on corrugated steel (popular for office buildings)

Beams

- timber (short spans only)
- in-situ reinforced concrete (general industrial/commercial)
- precast concrete (not common unless prestressed)
- prestressed concrete (suitable where long spans are required)
- steel

Columns

- timber (domestic and other small-scale construction only)
- reinforced concrete
- steel

Pitched roofs

- timber truss or rafter/purlin construction (domestic only)
- steel truss or portal frame (longer-spanned commercial/industrial buildings)

Foundations

- concrete (usually reinforced for other than domestic construction)

22 More on materials

Material selection for structural design

In earlier chapters of this book we have looked at such matters as shear force, bending moment and stress. We have learned how to evaluate these things and, in the case of shear force and bending moment, how to draw diagrams of their distribution. Some readers may have wondered how we apply this information. For example, we might calculate that the maximum bending moment experienced in a particular beam is 45 kN.m, or that the compressive stress in a certain column is 25 N/mm^2, but how do we make use of this knowledge?

The process of converting a piece of information such as maximum bending moment = 45 kN.m to a reinforced concrete or steel beam of a shape and size that will resist this bending moment is known as *structural design*. The full structural design process is beyond the scope of this book – there are many excellent textbooks available on the subject – but this chapter serves as an introduction to structural design.

The first decision that the structural designer needs to make is what material – or combination of materials – should be used in a given situation. In Chapter 21 we discussed the four main materials used in structural design (steel, reinforced concrete, masonry and timber), the advantages and disadvantages of each, and which material(s) is likely to be used for any particular type of structural member. This should guide you in your material selection. We will now discuss the alternative forms of building construction that are available to the designer.

Alternative forms of construction

The most common types of structural schemes for buildings are outlined in the following sections.

Steel frames

These are structural frameworks comprising steel beams and columns supporting floor slabs. The floor slabs are usually of concrete or a steel/concrete composite such as profiled steel decking onto which concrete is poured. The beams that span between columns are called *primary* beams and they in turn may support *secondary* beams (we saw an example of this in Chapter 5).

Basic Structures, Second Edition. Philip Garrison.
© 2011 John Wiley & Sons, Ltd. Published 2011 by John Wiley & Sons, Ltd.

Fig. 22.1   Steel-framed office building under construction.

Lateral stability of the structure is important and we saw in Chapter 11 that this may be assured by using diagonal cross-bracing or by designing the beam/column joints to be sufficiently rigid. Such measures may also be necessary to prevent torsion (twisting) of the building.

Steel beams and columns are available from manufacturers in standard section sizes, and tables of the structural properties of these standard sizes are available for designers. While deeper sections may be stronger and lighter (and hence, in material terms, cheaper) than shallower ones, headroom considerations may lead to the over-all building height being greater if deeper sections are used. This will lead to increased costs because the increased height of the building means that a greater number of columns, cladding, lifts, etc. is required.

Services (that is, electrical and telephone cables, gas and water pipes and ventila-tion ducts) need to be accommodated. We will see in Chapter 23 that some types of steel beam can cater for such services more readily than others.

Steel beams and columns need to be connected to each other, usually by bolting or welding. Connection details need to be kept simple in order to keep the costs (of fabrication, installation and material) to a minimum. And, as we've already seen, steel is vulnerable to fire and corrosion and needs to be protected accordingly. Because steel sections are slender, they are also vulnerable to buckling, a consideration that needs to be addressed at the design stage.

Figure 22.1 shows a typical steel-framed office building under construction.

Reinforced concrete frames

These are frames of concrete beams and columns. As we saw in Chapter 21, structural concrete is always reinforced internally with steel bars in order to provide the required tensile strength. Reinforced concrete frames usually comprise in-situ concrete: this means that the concrete beams or columns are formed by pouring wet concrete into a mould (formwork) located at the beam or column's final position. The formwork needs to be supported by a temporary propping structure which, along with the formwork itself, needs to be left in position for several days until the concrete has gained sufficient strength. This requirement can impede and delay other site activities.

Reinforced concrete beams support reinforced concrete slabs, which can be of various types, for example, one-way spanning, two-way spanning, rib or waffle, which were illustrated in Chapter 3. Unlike steel structures, fire protection is not normally a problem with concrete and, provided cracking is kept within acceptable limits, neither is corrosion of the steel reinforcement.

Construction of reinforced concrete buildings is quite labour-intensive: operatives are required to make and install the formwork and its supports, place the reinforcement and place the concrete.

Precast concrete frames

Precast concrete frames comprise individual beams and columns of reinforced concrete that have been made in a factory then delivered to site in their completed form. (This contrasts with the in-situ concrete frames discussed above, where the concrete is formed at its final position on site.) Greater quality control can normally be achieved with precast members, as the environment in a factory is more controllable than that on site. Also, as the precast members will have achieved their full strength when they arrive on site, there will be no waiting time to delay other site activities. However, precast construction is best suited to structures which are totally regular and repetitive in nature.

Timber frames

Because of timber's limited strength, timber-framed structures tend to be small-scale domestic buildings. Timber roof frames are generally used for domestic pitched roofs. Connection between members, and protection of timber against rot or insect attack, is also a consideration.

Load-bearing masonry

Structural masonry – stone, brickwork and blockwork – is the preferred form of construction for the walls of houses and other non-framed structures in the UK that and elsewhere. Masonry's high compressive strength makes it ideal for such structures and also for other structures that are in pure compression: for example, arches. Masonry comes in small units (e.g. individual bricks) which are easy to manage. However, skilled labour is required.

Masonry is generally less tolerant to differential settlements and accidental damage than steel or in-situ concrete-framed buildings.

Hybrid schemes, e.g. steel frame with precast concrete floors

These are combinations of the above.

The choice between different construction types

The choice between construction types will depend on the following factors.

The need for flexibility

As we saw in Chapter 21, the future use of the building – and whether this is likely to change over time – may influence the type of construction.

The spans required

Sometimes there are requirements for long uninterrupted spans in, for example, theatres and other auditoria, multi-storey car parks, exhibition halls or, in the case of bridges, shipping lanes. In such cases the span will usually dictate the form of construction. As a general rule, the longer the crossing, the more expensive it will be to achieve. See Chapter 23 for more about this.

The ground conditions

These will dictate whether relatively cheap conventional foundations can be used or whether more expensive piling or other foundation types are required.

The access to the site

If the site is isolated, access roads may need to be built. Then again, if the site is in the middle of a large city, delivery of materials may be restricted. There may be other constraints, for example, all materials may have to be lifted in by crane because of physical site constraints. The wise designer considers all this at the design stage.

The experience of the designers

The designers may have great experience of a particular type of design which can be used on the present project. This makes the whole process less painful – and less costly – because of the benefits of the lessons learned on previous projects.

The experience of the contractors

Again, the contractors may have experience of certain construction types and techniques which, if used on the present project, will keep the cost down.

The availability of materials

It is no use specifying materials which are either unavailable or have to be imported at great cost, whatever their attributes might be.

Risks and difficulties in the construction process tend to lead to increased costs, so the chosen solution should seek to minimise these. Once you have decided on the material (or combination of materials) to use, the design process can begin.

Design from first principles and design standards

You can learn how to design (for example) a reinforced concrete beam from first principles. This largely mathematical process is taught at some universities and is dealt with in some textbooks. But bear in mind that you aren't the first person to attempt to design a reinforced concrete beam and you won't be the last. The problems that you encounter in doing so have all been encountered before, and attempts have been made to deal with the process in documents called *standards* or *codes of practice*. In the remainder of this discussion I will assume that you are either in the UK or some other places where British Standards are used, as my discussion will be based on British Standards. However, the general principles of what I say will also apply to other standards (e.g. the American ASTM Standards). In addition, you should be familiar with the local building codes of the country or place where you are, as local building codes overrule the requirements of standards.

British Standards

Are British Standards a set of rules and regulations, legal documents or design guides? Certainly, all design must conform to the requirements of the relevant British Standard, but to what extent do British Standards fulfil the role of design guides? Certainly, a person who is given half an hour to design a reinforced beam is unlikely to find the relevant British Standard much use unless he or she has been trained in putting it into practice. However, a person familiar with the standard will be able to use it to design a beam very quickly.

The early British Standards may have been user-friendly design guides. However, British Standards have become more and more voluminous over the years and, in my opinion, now read more like legal documents.

An individual new to structural design needs guidance in the use of British Standards. If you study a course or module entitled 'structural design' (or similar) at university or college, your lecturer or tutor will (or should) act as a friendly guide through the relevant parts of the code. Some textbooks are also good at performing this role.

The good news is that the design of, say, a timber beam or a structural steelwork column is a fixed procedure which can be easily followed or learned – effectively, once you've designed one masonry wall you've designed them all!

The relevant British Standards and where to find further information

The relevant British Standards are:

- BS 8110: Part 1: 1997: Structural use of concrete.
- BS 5628: Part 1: 1992: Structural use of unreinforced masonry.
- BS 5268: Part 2: 1996: Structural use of timber.
- BS 5950: Part 1: 2000: Structural use of steelwork in building.

I'm sure it's totally unintentional on the part of the drafters of the British Standards, but you'll realise how easy it is to confuse the masonry and timber standards with each other (BS 5628 and BS 5268 respectively).

Officially, Eurocodes replaced British Standards in 2010, though old habits die hard and it is likely that British Standards will still be in use for some time to come, not least because of their familiarity and relative simplicity.

Eurocodes are used for design throughout the European Union and it is hoped that this international consistency will make it easier for engineers from different countries to work together. Each Eurocode should be read in conjunction with the relevant National Application Document (NAD) which will give parameters which should be used for specific countries. The relevant Eurocodes are listed below:

- EC0 Basis of structural design
- EC1 Actions on structures
- EC2 Design of concrete structures
- EC3 Design of steel structures
- EC4 Design of composite steel and concrete structures
- EC5 Design of timber structures
- EC6 Design of masonry structures
- EC7 Geotechnical design
- EC8 Design of structures for earthquake resistance
- EC9 Design of aluminium structures

Apart from the first, each of these Eurocodes is divided into a number of parts.

23 How far can I span?

Introduction

Any person involved in the conceptual design of a building will soon have to consider how far can be spanned in practice. There are no easy answers to this question (although some rules of thumb are given below), but this chapter explores the various factors involved.

Long span structures

What is a span?

You will be most familiar with the word *span* in the context of bridges, but it applies to beams and slabs within building structures as well. The span of a bridge (or beam, or whatever) is the horizontal distance between supports.

How far can we span?

On the face of it, a span can be as long as is necessary. A concrete lintel struggles to span a 1 metre wide door opening, while modern suspension bridges can – and do – span several kilometres. In practice, the greater the span, the stronger the spanning element has to be. This generally means that it has to be deeper, and this greater depth may be difficult to accommodate physically. And inevitably, as span increases, the cost will also increase.

However, spans should not be too small either, as an excessive number of columns or supporting walls can interfere too much with activities going on within the building – nobody wants to see a 'forest' of columns. In general, spans are made as long as is reasonably practical.

Some points are worth noting:

- Some materials are stronger than others, therefore can span greater distances.
- In general, the longer the span, the deeper the supporting element has to be.
- Some building uses dictate that large uninterrupted spans are required. Examples include sports halls, swimming pools, theatres and concert halls.

Basic Structures, Second Edition. Philip Garrison.
© 2011 John Wiley & Sons, Ltd. Published 2011 by John Wiley & Sons, Ltd.

Fig. 23.1   **Building with cantilevered canopy, Berlin.**

Fig. 23.2   **Building with cantilevered canopy, Berlin (detail).**

At first glance, the canopy shown in Fig. 23.1 looks impossibly wafer-thin in respect of the distance it is cantilevering. However, Fig. 23.2 shows a break in the canopy which reveals that it is in fact considerably deeper (and hence stronger) at its supporting columns. Careful design conceals this fact elsewhere from street-level observers.

Let's consider the various structural materials one by one.

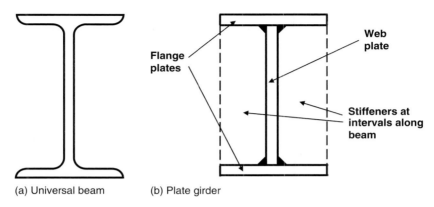

Fig. 23.3   **Universal beams and plate girders.**

Steel

Standard steel beams

Steel beams (universal beams) are manufactured in standard sizes. Corus (formerly British Steel, and now owned by Tata Steel) produces tables of these standard sizes and their various dimensions and properties. In many cases a steel building will comprise standard beams (and columns), selected by the designer to accommodate the calculated bending moments, shear forces and deflections. A typical universal beam section is shown in Fig. 23.3a.

Plate girders

The largest standard section Tata produces is 914 mm deep. In situations where even this largest size is not adequate, it is possible to fabricate larger sections by welding together plates in the right configuration. In other words, a section can be 'tailor made' when no off-the-peg section is adequate. A typical plate girder section is shown in Fig. 23.3b.

Castellated and cellular beams

These steel beams have large holes in their webs (that is, the vertical parts) at regular intervals along the beam. These holes are hexagonal in the case of castellated beams and circular in the case of cellular beams. Cellular beams in particular are very popular in multi-storey steel-framed construction. Castellated beams are formed by cutting a standard steel beam longitudinally along a zigzag, as shown in Fig. 23.4a, then reconnecting the two half beams as shown in Fig. 23.4b, thus forming a deeper (and therefore stronger) section of the same weight as before. Moreover, the holes in castellated and cellular beams can also be used to accommodate services such as cables or water pipes.

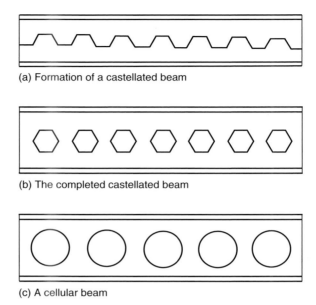

(a) Formation of a castellated beam

(b) The completed castellated beam

(c) A cellular beam

Fig. 23.4   Castellated and cellular beams.

Lattice girders

But what happens when a plate girder section would have to be so big as to be impractical? Well, instead of using a solid steel beam (as shown in elevation in Fig. 23.5a), a lattice girder beam could be used. As you will recall, the top of a sagging beam is in compression and the bottom is in tension. A lattice girder comprises a top boom (in compression) and a bottom boom (taking the tension), with the two booms linked by diagonal members. Using a lattice girder, a deep beam can be achieved without the requirement for it to be solid. This saves on material (and thus weight) and means that the gaps within the lattice can be used for other things (e.g. services can pass through them). A typical lattice girder is shown in Fig. 23.5b.

Lattice girders (referred to there as 'bar joists') are a common type of floor construction in commercial buildings in North America. These bar joists are typically 300–400 mm deep and are spaced at (typically) 600 mm.

Lattice box girders

What happens when the required span and loading increase still further? We could continue deepening (and thus strengthening) the lattice girder. Another option is to introduce a second lattice girder running alongside the first one and linked to it by two horizontal lattice girders, one at top boom level, the other at bottom boom level. A box is thus formed and therefore this type of beam is called a box lattice truss, shown in Fig. 23.6a. A variation on this theme is the triangular lattice truss, shown in Fig. 23.6b. As steel is more likely to buckle when in compression, the cross section of steel available in the compression zone is maximised by having two booms in this zone and one in the

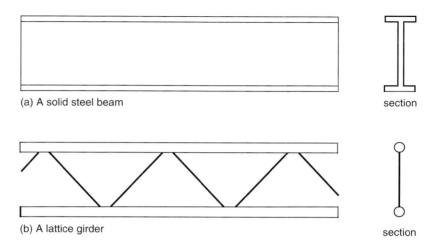

(a) A solid steel beam   section

(b) A lattice girder   section

Fig. 23.5   Solid steel beams and lattice girders.

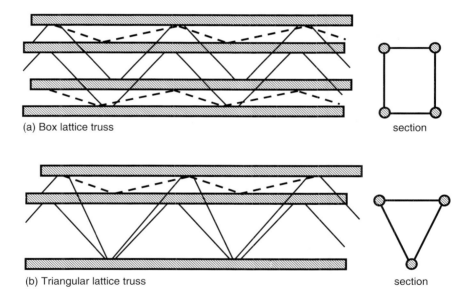

(a) Box lattice truss   section

(b) Triangular lattice truss   section

Fig. 23.6   Lattice trusses.

bottom (tensile) zone as shown. Figure 23.7 shows triangular steel lattice trusses, double curvature in profile, spanning an airport concourse and supporting its roof.

Suspension structures

Lattice box girders can span considerable distances – typically up to 100 metres when supporting football stands – but there may be cases when we need to span greater distances. In such cases we need to use suspension or cable-stayed structures. The

Fig. 23.7   Triangular lattice trusses supporting terminal roof, Liverpool John Lennon Airport.

principle behind these is that if support cannot be provided from below, it can be provided from above by means of cables which run over supporting masts to an anchorage point in the ground.

Concrete

Reinforced concrete

As discussed in Chapter 21, structural concrete is always reinforced – i.e. it has steel bars embedded in it – for strength purposes. Unfortunately, the span/depth ratios required of reinforced concrete are not very desirable for designers: if a reinforced concrete beam is required to span a long distance, its depth will be inconveniently great.

Prestressed concrete

Prestressed concrete beams (i.e. those containing embedded steel bars or cables subjected to large tensile forces) can be much more slender than reinforced concrete beams and therefore are a popular choice when long spans are required. Prestressed concrete beams are often visible in multi-storey car parks.

Timber

Timber beams and joists

The material aspects of timber were discussed in Chapter 21, where we saw that timber beams cannot span great distances because of their limited strength. The cross-sectional size is also limited by the size of the tree from which the timber was obtained.

Glued-laminated ('glulam') beams

Longer spans are possible with timber if glued laminated beams are used. As mentioned in Chapter 21, such beams are formed by building up layers from thin slices of timber glued together.

Glulam beams are not common in the UK due to their high cost, but are sometimes seen supporting the roofs of swimming pools – timber being less susceptible to the corrosive action of chlorine gas than other materials.

Masonry

As discussed in Chapter 21, masonry is weak in tension and is therefore not really suitable as a spanning material. This is why the columns in ancient Greek and Egyptian temples are so close together: the stone beams that span between them can span only a short distance.

Table 23.1   Span ranges and span/depth ratios.

Type of element	Span range	Typical span/depth ratio
Concrete		
Beam, simply supported	Up to 8 m	15–20
Beam, continuous	Up to 12 m	20–27
Beam, cantilever	Up to 5 m	1–7
Slab, one-way, simply supported	Up to 6 m	20–30
Slab, one-way, continuous	Up to 6 m	20–30
Slab, one-way, cantilever	Up to 3 m	5–11
Slab, two way, simply supported	Up to 6 m	30–35
Slab, two way, continuous	Up to 6 m	30–35
Profiled steel decking/ concrete composite	Up to 6 m	35–40
Ribbed slab	Up to 11 m	35–40
Waffle slab	Up to 15 m	18–25
Column	Storey height	10–17
Strip foundation	0.8–2.0 m wide	
Pad foundation	1.5–3.0 m square	
Timber		
Joist flooring	Up to 6 m	10–20
Glulam beam	Up to 30 m	15–20
Ply-web beam	Up to 20 m	10–15
Steel		
Primary beams (supported by columns)	Up to 12 m	15–20
Secondary beams (supported by other beams)	Up to 7 m	15–20
Portal frame	Up to 60 m	35–40

Masonry arch structures

Masonry, being strong in compression, is suitable for use in arch structures, which are in compression throughout. Masonry arches can span reasonable distances, and a series of masonry arches can be used to form a viaduct, as can be seen in Roman stone aqueduct structures (for example, at Nimes in the south of France) and in Victorian brick railway viaducts at many locations in the UK and elsewhere.

Spans and depths: some rules of thumb

A question I am commonly asked by students of architecture is: 'How far can I span and how deep would the beam have to be?' If only it were that simple.

As mentioned earlier, in broad terms, the greater the span, the greater the depth. It follows that rule of thumb span-to-depth ratios can be generated and these are given in Table 23.1. These should be used with caution and the following points should be noted:

- The possible spans, and associated depths, depend on the loading to which the beam is subjected. The figures in Table 23.1 assume 'normal' commercial building loads. They do not apply to more heavily loaded situations (e.g. plant rooms) or to unconventional loading scenarios.
- This information is given without prejudice and is for guidance purposes only. It is suitable for initial sizing of structural elements for architectural scheme or costing purposes.
- For actual building projects the size of structural elements must be verified through detailed design by a qualified structural engineer.

24 Calculating those loads

Introduction

In the earlier chapters of this book you were shown how to calculate such things as shear force, bending moment and stresses. In worked examples you were presented with loads to work with. Unfortunately, real-life structural problems are not as neatly packaged as examples you might encounter in textbooks or in university lecture theatres. You have to calculate the loads yourself. This chapter tells you how.

As you learned in Chapter 5, there are two types of loading:

1) dead (or permanent) loads;
2) live (or imposed) loads.

We will discuss the calculation of each of these in turn, then look at some examples.

Dead load

Unit weights of common building materials are given in Appendix 1. (British Standard BS 648 gives the unit weights of a much wider range of materials.) These loads are expressed in kN/m^3 and represent the weight of a cubic metre of the material. For example, the unit weight of reinforced concrete is 24 kN/m^3, which means that a cubic metre of concrete weighs 24 kN. This is almost two-and-a-half times the weight of water. So if, in the early stages of your career, you have to carry buckets full of wet concrete short distances on construction sites (as I did), you'll find they're considerably heavier than the buckets of water you use when cleaning your car!

If you want to calculate the weight of a reinforced concrete beam which is 200 mm (or 0.2 metres) wide, 400 mm (or 0.4 metres) deep and 6 metres long, you first need to calculate the volume of the beam, then multiply it by the unit weight to get the total weight.

$$\text{Volume of beam} = \text{length} \times \text{breadth} \times \text{height}$$
$$= 6\,\text{m} \times 0.2 \times \text{m} \times 0.4\,\text{m}$$
$$= 0.48\,\text{m}^3$$

$$\text{Total weight of beam} = \text{volume} \times \text{unit weight}$$
$$= 0.48\,\text{m}^3 \times 24\,\text{kN}/\text{m}^3$$
$$= 11.52\,\text{kN}.$$

Basic Structures, Second Edition. Philip Garrison.
© 2011 John Wiley & Sons, Ltd. Published 2011 by John Wiley & Sons, Ltd.

Live load

As you will recall from Chapter 5, these are loads due to people and furniture. By their very nature, they are variable. To simplify matters, live loads are assigned certain values depending on the use of the building or area concerned. These loads are expressed in kN/m^2 and typically fall in the range $1.5–5.0 \, kN/m^2$. Values for some common cases are given in Appendix 1.

For example, the live load relevant to classrooms is $3.0 \, kN/m^2$. So, for a classroom which has a floor area 10 metres × 10 metres:

$$\text{Total live load} = 10 \, \text{m} \times 10 \, \text{m} \times 3.0 \, kN / m^2$$
$$= 300 \, kN.$$

Example 24.1: Loading on a reinforced concrete beam

A reinforced concrete beam spans 6 metres between supporting columns. The beam is 250 mm wide and 450 mm deep and supports a 5 metre wide portion of slab 175 mm deep. There is a 40 mm deep concrete topping layer on top of the slab. The floor supports offices. There is also a non-load-bearing masonry (blockwork) wall directly above the beam and running along the line of the beam. This blockwork wall is 2.5 metres high, 200 mm thick and is finished on both sides with plaster of weight $0.5 \, kN/m^2$. See Fig. 24.1.

Calculate the total load on the concrete beam per metre length. (Note: don't forget to include the weight of the beam itself.)

Solution

Unit weight of concrete = $24 \, kN/m^3$ (Appendix 1)
Unit weight of blockwork = $22 \, kN/m^3$ (Appendix 1)
Live load due to offices = $2.5 \, kN/m^2$ (Appendix 1)

Note that you are asked to calculate the total load per metre length of concrete beam. There will be a number of contributions to this total load. They come from the blockwork wall, the plaster on it, the topping layer on top of the slab, the slab itself, the beam's own weight and the live load (due to people and furniture) on the slab. One of the basic mistakes students sometimes make with this sort of calculation is in forgetting one or more of these contributions.

Considering these contributions one by one:

Blockwork wall:

This is simply a matter of multiplying the volume of the wall (length × breadth × height) by its unit weight:

$$\text{Load due to blockwork wall} = 2.5 \, \text{m} \times 0.2 \, \text{m} \times 1.0 \, \text{m} \times 22 \, kN/m^3$$
$$= 11 \, kN$$

Plaster on blockwork wall:

The unit weight of plaster has been expressed in units of kN/m^2 – in other words, load per unit area. This means that we must calculate the total plastered area per metre

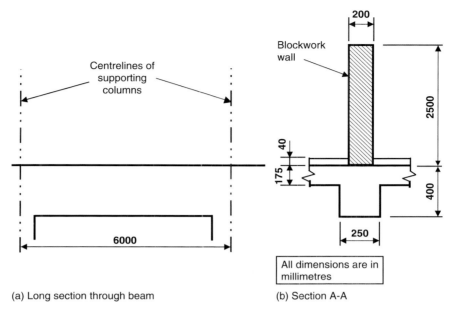

(a) Long section through beam   (b) Section A-A

Fig. 24.1   Concrete beam featured in example 24.1.

length of wall and multiply this area by the unit weight of plaster given above. Remember, the wall is plastered on both sides, so the number 2 in the calculation below represents two sides:

Load due to plaster $= 2 \times 2.5\,\text{m} \times 1\,\text{m} \times 0.5\,\text{kN/m}^2 = 2.5\,\text{kN}$

Concrete slab:

As with the blockwork wall, we multiply the volume of the slab (per metre length of beam) by the unit weight of reinforced concrete. Remember: the beam is supporting a 5 metre wide portion of slab.

Load due to reinforced concrete slab $= 5\,\text{m} \times 0.175\,\text{m} \times 1\,\text{m} \times 24\,\text{kN/m}^3$
$= 21\,\text{kN}$

Topping layer:

It is reasonable to assume that the unit weight of the topping layer is the same as that for the structural reinforced concrete. As with the slab proper, we multiply the volume of the topping by the unit weight. To make calculations easier, we will pretend that the topping continues underneath the blockwork wall – even though it doesn't. This means our calculation will be slightly conservative (i.e. an over-estimate of the total force).

Load due to topping layer $= 5\,\text{m} \times 0.04\,\text{m} \times 1\,\text{m} \times 24\,\text{kN/m}^3 = 4.8\,\text{kN}$

Concrete beam:
We have already considered the top part of the beam (i.e. the top 175 millimetres of the beam's depth) when we were calculating the load due to the slab. We now need to calculate the loading due to the bottom 225 mm (i.e. 400 − 175) of the beam.

$$\text{Load due to reinforced concrete beam} = 0.175\,m \times 0.250\,m \times 1\,m \times 24\,kN/m^3$$
$$= 1.05\,kN$$

Live load:
This was expressed above as a load per unit area of floor slab ($2.5\,kN/m^2$). Again, for simplicity we'll ignore the presence of the blockwork wall when calculating this load. The live load will be the surface area of the slab multiplied by the load per unit area, as follows:

$$\text{Live load} = 5\,m \times 1\,m \times 2.5\,kN/m^2 = 12.5\,kN$$

So:

$$\text{Total dead load} = (11 + 2.5 + 21 + 4.8 + 1.05) = 40.4\,kN$$
$$\text{Total live load} = 12.5\,kN$$

This gives a total load of 52.9 kN per metre length of the beam.

Example 24.2: Loading at the base of a column

A four-storey reinforced concrete-framed building has a plan area of 18 metres × 25 metres. Supporting columns are arranged on a grid of 6 metres × 5 metres, as shown in Fig. 24.2. At each level, floor slabs span 5 metres onto supporting beams, which in turn span 6 metres between columns. The 6 metre span beam considered in Example 24.1 is a typical supporting beam.

If the ground floor slab is ground-bearing (in other words, it is supported directly on the ground underneath) and the live load on the flat roof is the same as on the floors, calculate the total load at the base of a typical internal supporting column if the columns are 400 mm × 400 mm in plan.

Solution

At each level, a typical column will support an area of beam and slab as shown by the hatched zone in Fig. 24.2. In Example 24.1 we have already calculated the total load on a typical 1 metre length of beam, so if we multiply this figure by 6 metres, we have the total load supported by a typical column at each level. We then need to multiply the result by 4 to represent the four floors (excluding the ground-bearing ground floor slab but including the roof slab).

$$\text{Total load on typical column from beams} = 52.9\,kN/m \times 6\,m \times 4\,storeys$$
$$= 1270\,kN$$

We now need to add on the weight of the column. If the total height of the building (and thus of the column) is 14 metres, the weight of the column is:

$$\text{Column self-weight} = 14\,m \times 0.4\,m \times 0.4\,m \times 24\,kN/m^2 = 53.8\,kN$$

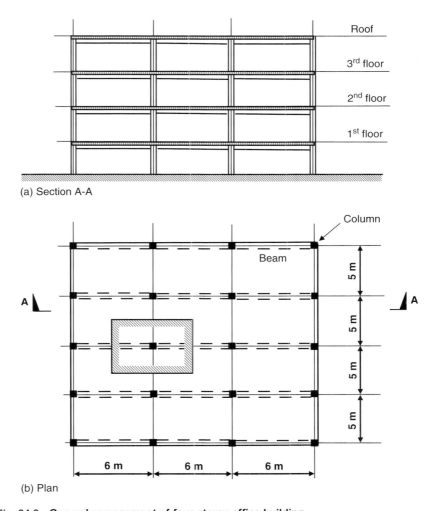

(a) Section A-A

(b) Plan

Fig. 24.2   General arrangement of four-storey office building.

(Again, this is obtained by working out the volume of the column and multiplying it by the unit weight of concrete.)
So:

Total load at base of a typical internal column is: $1270 + 53.8 = 1323.8\,kN$

Example 24.3: Sizing of a pad foundation

Typically, a column in a building will be supported by a pad foundation. The function of a pad foundation – indeed, any foundation type – is to transmit the loads from the building's superstructure (that is, the above ground part) safely into the ground beneath.

In order to determine a foundation size, two things need to be known:

- the total load on the foundation
- the permissible ground-bearing pressure

The permissible ground-bearing pressure – in other words, the maximum pressure that the ground can sustain without deforming – can only be determined from a ground investigation relating to the site of the proposed building.

If a ground investigation has been done for the site of the building discussed in Example 24.2 and the permissible ground-bearing pressure has been found to be $200\,\text{kN/m}^2$, calculate the pad foundation size required.

Solution

$$\text{Minimum pad size required} = \frac{\text{Total column load}}{\text{Permissible ground bearing pressure}}$$

$$\text{Minimum pad size required} = \frac{1323.8\ \text{kN}}{200\ \text{kN/m}^2} = 6.62\ \text{m}^2$$

Normally, pad foundations are square, except when practical constraints – for example, the presence of obstructions – mean that they have to be rectangular. So, if a square pad has to have a minimum plan area of $6.62\,\text{m}^2$, the minimum length of one of its sides is the square root of 6.62, i.e. 2.57 metres.

We shall round up this value to 2.7 metres, for the following reason. The self-weight of the base also acts on the ground below of course. But we cannot calculate the weight of the base until we know its size. By rounding up the base side length to 2.7 metres, we are increasing the base size to allow for the extra load due to the base itself. Let's assume that the depth of the foundation is 0.5 metres. We then need to check that the actual ground-bearing pressure is less than $200\,\text{kN/m}^2$:

Total load in column = 1323.8 kN (calculated above)
Weight of 2.7 m square base = 24 kN/m3 × 0.5 m × 2.7 m × 2.7 m
= 87.5 kN

Total load = 1323.8 + 87.5 = 1411.3 kN

So:

$$\text{Actual ground bearing pressure} = \frac{\text{Load}}{\text{Base area}} = \frac{1411.3\ \text{kN}}{2.7\ \text{m} \times 2.7\ \text{m}}$$

$$= 193.6\ \text{kN/m2}$$

As this is less than $200\,\text{kN/m}^2$, a pad foundation of plan dimensions 2.7 m × 2.7 m is satisfactory.

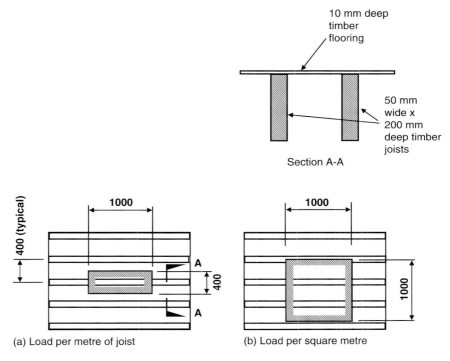

Section A-A

10 mm deep timber flooring

50 mm wide x 200 mm deep timber joists

400 (typical)

1000

400

A

A

(a) Load per metre of joist

1000

1000

(b) Load per square metre

Fig. 24.3   Loads on timber joist flooring.

Example 24.4: Loads in timber joist flooring

Due to its relatively low strength, timber flooring tends to be used in domestic construction where loading is light and spans are comparatively short. Timber flooring comprises timber beams (or joists, as they are usually known) at fairly close centres (typically 400, 450 or 600 mm).

For example, if a timber floor comprises 50 × 200 (width × depth in millimetres) timber joists, spaced 400 mm apart, supporting 10 mm thick timber boarding, the load on every metre length of joist (see Fig. 24.3a) is as calculated below. Assume an imposed load of 1.5 kN/m² (normal in domestic construction) and that the unit weight of softwood is 5.9 kN/m³ (see Appendix 1).

$$\text{Self-weight of joist per metre length} = (0.05\,\text{m} \times 0.2\,\text{m} \times 1.0\,\text{m} \times 5.9\,\text{kN/m}^3)$$
$$= 0.059\,\text{kN}$$

$$\text{Self-weight of boarding per metre of joist} = (1.0\,\text{m} \times 0.4\,\text{m} \times 0.01\,\text{m} \times 5.9\,\text{kN/m}^3)$$
$$= 0.024\,\text{kN}$$

$$\text{Live load per metre length of joist} = (1.0\,\text{m} \times 0.4\,\text{m} \times 1.5\,\text{kN/m}^2)$$
$$= 0.6\,\text{kN}$$

So:

$$\text{Total load per metre length of joist} = (0.059 + 0.024 + 0.6) = 0.683\,\text{kN}$$

Now let's suppose we wanted to calculate the load per square metre of flooring. A square metre of flooring with joists at 400 mm spacing will contain 1/0.4 = 2.5 metres of timber joist (see Fig. 24.3b). The total load per square metre of flooring is calculated as follows:

$$\text{Self-weight of 2.5 m length of joist} = (0.05\,\text{m} \times 0.2\,\text{m} \times 2.5\,\text{m} \times 5.9\,\text{kN/m}^3)$$
$$= 0.148\,\text{kN}$$

$$\text{Self-weight of one square metre of boarding} = (1.0\,\text{m} \times 1.0\,\text{m} \times 0.01\,\text{m} \times 5.9\,\text{kN/m}^3)$$
$$= 0.059\,\text{kN}$$

$$\text{Live load on one square metre of boarding} = 1.5\,\text{kN}$$

Therefore

$$\text{Total load per square metre of timber flooring} = (0.148 + 0.059 + 1.5) = 1.71\,\text{kN}$$

From inspection it can be seen that the dead load for timber flooring is usually a lot less than the associated live load.

The dead load part of the above calculation is $(0.148 + 0.059) = 0.207\,\text{kN/m}^2$.

In general, a total dead load of $0.25\,\text{kN/m}^2$ is a convenient figure to use when performing calculations involving timber flooring.

Example 24.5: Loads due to steel beams

Structural steel beams come in standard sizes, each of which is designated by three figures multiplied together. The first figure is the nominal width, the second figure is the nominal overall depth and the third figure is the steel beam's own weight expressed in kg per metre length.

For example, the universal beam section designated 203 × 133 × 23 has a depth of approximately 203 mm, a width of around 133 mm and each metre of it weighs 23 kg.

Therefore, if we know the specific steel beam that is being used in a given situation, we know the self-weight in kg/m – which can be converted to kN/m by dividing by 100. For example, 23 kg/m = 0.23 kN/m.

If we don't know the specific steel beam size that is being used in a given situation, we can estimate the self-weight using the following rules of thumb:

- For steel beams up to 360 mm deep, self-weight (in kg/m) is about one-sixth of the depth (e.g. a 203 mm deep beam weighs 203/6 = 34 kg/m).
- For steel beams 360 to 800 mm deep, self-weight (in kg/m) is about one-quarter of the depth (e.g. a 533 mm deep beam weighs 533/4 = 133 kg/m).
- For steel beams over 800 mm deep, self-weight (in kg/m) is about one-half of the depth (e.g. a 914 mm deep beam weighs 914/2 = 457 kg/m).

Example 24.6: Loads on the supports of timber joist flooring

A timber floor comprises 50 × 200 timber joists spanning 4 metres from a central supporting beam to load-bearing walls on either side, as shown in Fig. 24.4. The

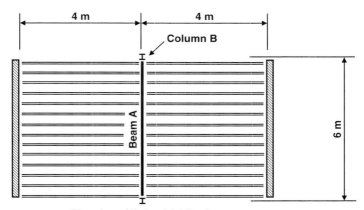

Plan view of timber joist flooring

Fig. 24.4   Loads on supports of timber joist flooring.

central supporting beam, labelled 'Beam A', is a 203 mm deep steel beam and is supported by steel columns – labelled 'Column B' – at each end of its 6.0 metre span. If the dead load of the timber flooring is 0.25 kN/m² and the live load is 1.5 kN/m², calculate:

● the total load per metre length of beam A
● the total load on each of the two columns B.

As the span of the flooring is 4.0 metres, half of that span (i.e. 2 metres) is supported by beam B. But it is 2 metres *on each side* of the beam, so the total portion of flooring supported by the beam = 2 × 2 m = 4 metres.

Total load per square metre of flooring = (0.25 + 1.5) = 1.75 kN/m²
Estimated self-weight of 203 mm deep steel beam (see example 24.5 above)
$$= 203/6 = 34 \, \text{kg/m}$$
$$= 0.34 \, \text{kN/m}$$

Total load on beam B (per metre) from flooring = (4 m × 1 m × 1.75 kN/m²)
$$= 7.0 \, \text{kN}$$
Self-weight of beam B per metre = 0.34 kN

Therefore

Total load per metre length of beam B = (7.0 + 0.34) = 7.34 kN
$$\text{Total load on each supporting column} = \frac{7.34 \, \text{kN} \times 6 \, \text{m}}{2} = 22 \, \text{kN}$$

Fig. 24.5   Atrium, Learning Centre, Leeds Metropolitan University.

Fig. 24.6   Office building, Deansgate, Manchester.

Figure 24.5 shows an atrium. Atriums are becoming increasingly common and feature large areas of glass (which may be horizontal, vertical or inclined). The glass needs to be supported and the supporting structure may be substantial.

Figure 24.6 shows the base of a modern high-rise office building. Note the inclined supports.

25 An introduction to structural design

General

The great thing about studying civil and structural engineering is that examples are all around you, particularly in a large modern city. The whole world is a real-life full-scale civil and structural engineering laboratory. If travelling by train, for example, or by car along a motorway, you'll see different styles of bridges, and can reflect on why a particular bridge type was chosen for any given location. Walking through a town or city, you'll encounter many different building types. As we've discussed, no one building material is ideal for every situation, so you'll encounter a mix of building materials, as well as a mix of building ages. Whenever you've time to spare in a built-up area – for example, while waiting at a bus stop, or having a drink in an outside cafe or bar – take time to study the buildings around you; try to work out why that particular material and form of construction was chosen, think about any constructional problems that might have been experienced and ponder on how that building relates to other buildings around it. Discuss it with your friends, and bore your relatives – well, why not?!

You'll normally study structural design on Level 2 of your course, and this chapter serves as an introduction to that process. There are many textbooks on structural design. Some discuss design as a concept, with little regard to the actual design process for real buildings, others swamp the reader with reams of calculations of varying degrees of complexity and comprehensibility. In this chapter, I aim to steer a middle course: I'll be talking about the process of structural design, telling you how it's done, at the same time keeping away from calculations and keeping the amount of symbols and jargon to a minimum. In practice, structural design is carried out using design standards, which in the UK are British Standards and Eurocodes (as mentioned in Chapter 22), but I'm trying to keep the discussion general so I shall avoid reference to specific standards as much as possible.

If you've read the preceding part of this book, you'll have recognised that common structural elements behave in certain ways. For example, beams and slabs span horizontally between supports and experience bending, and we've seen in previous chapters that this leads to tension in one of the faces (perhaps the bottom face) and compression in the opposite face (perhaps the top). Columns, which are vertical support 'pillars', experience axial compression in simple cases, but if the load on the column is eccentric then the axial compression is combined with bending about one or both axes. Foundations, by contrast, have not hitherto been discussed in any depth

Basic Structures, Second Edition. Philip Garrison.
© 2011 John Wiley & Sons, Ltd. Published 2011 by John Wiley & Sons, Ltd.

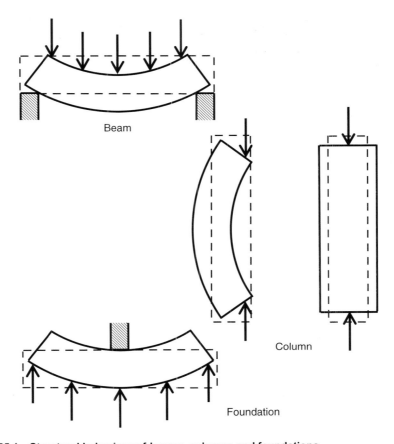

Fig. 25.1   Structural behaviour of beams, columns and foundations.

in this book, but they normally experience bending, as will be discussed below. See Fig. 25.1 for some sketches.

Structural design involves determining loads, then analysing the forces and stresses – which may be tensile, compressive, shear or bending – and determining the size of structural element in a given material which can cope with these forces and stresses.

Choice of material

Previous chapters have given guidance on choice of material. In framed structures, the choice is usually between steel and reinforced concrete.

Typical member sizes

In some cases below I've given a 'typical' size of beam (or column, or whatever) in that particular material. Where given, this is for general information and guidance only; there is no implication that that particular size would be appropriate in all (or indeed any given) situations.

A word about sustainability

This is a book about structures, and I'm not going to get involved in the current debate on global warming/climate change. However, it makes sense to conserve the earth's resources as much as possible, and structural designers – like everyone else – should be conscious of sustainability and should design their structures to be as sustainable as possible.

According to one definition, sustainable development meets the needs of the present without compromising the ability of future developments to meet their own needs. The implication of this is that damage to the environment and depletion of resources should be minimised.

Other ways of improving sustainability include:

- reducing energy use (in both construction and operation of buildings)
- extending a building's useful life
- reusing components and recycling materials when they are no longer needed
- sourcing materials in such a way that environmental impacts are minimised
- reducing demolition waste
- making good use of other waste products.

Two specific examples of how sustainability in structural design could be improved are using long spans and avoiding downstand beams.

Long spans

Long spans create much wider spaces between columns that can be used more flexibly. In other words, the relative absence of columns gives greater scope for any future adaptation of the building for a different use.

Downstand beams

The presence of downstand beams impedes the flow of air as part of any natural cooling system in the building. Flat soffits help in this regard and also facilitate the installation and refit of services.

Loads

We talked about the different types of load (or actions, in Eurocode-speak) in Chapter 5 of this book, and some load calculations are given in Chapter 24. If you are unsure of the meanings of the terms dead load (or permanent load) and live load (or imposed load), you should return to those chapters and read all about it.

Factors of safety

Suppose you've been asked to design a floor slab that will support a school classroom that, when full, will accommodate 50 students. You could assess the load due to 50 students (plus one teacher, and possibly a visitor). You would also need to assess the loads due to furniture and the weight of the floor slab itself, plus finishes. Hopefully

Table 25.1   Standard values of factors of safety for loads.

	Dead Load	Live Load
British Standard	1.4	1.6
Eurocode	1.35	1.5

you, as a qualified engineer, could produce a competent design, but you don't want your design to be so 'tight' that if a 51st student were to walk in to the finished class-room the floor would collapse. In order to allow for uncertainties of this nature, the loading is deliberately overestimated by multiplying by factors of safety which are approximately 1.5. See Table 25.1 for actual values used in certain design standards. Safety factors are also applied to material strength to allow for uncertainties in the properties and qualities of materials; a design will clearly be safer if the strength of the materials has been underestimated, in the same way as the design will be safer if the loads are overestimated.

Load combinations

The maximum load is not necessarily the worst load case. This is a concept often misunderstood by students. The following two examples might help.

Example 1: Timber bookshelf

Imagine a wooden plank used as a bookshelf. The shelf is supported at its centre and its two ends by metal brackets, which are secured to the wall.

As you know from experience, books are heavy. If the book is loaded along its entire length by heavy large format books, the wooden bookshelf may well deflect under the loading, in the manner shown in Fig. 25.2a. (Note that the deflection has been grossly exaggerated for clarity.)

If the books are now removed from the right hand half of the bookshelf – that is, the right-hand span between supporting brackets, the deflection of the bookshelf will now be as indicated in Fig. 25.2b.

Comparing Figs 25.2a and 25.2b it can be seen that the deflection is greater in the left-hand span when only that span is loaded. In other words, the worst case occurs when the loading is not the maximum possible.

Example 2: Man standing on a garden wall

A man (live load) stands on a garden wall (dead load). It's a windy day, and the wind (wind load) is blowing horizontally against the wall and the man.

Fig 25.3 shows the lines of action of these various loads, which produce overturn-ing and restoring moments about the front edge of the base of the wall. The wind is trying to blow the wall over, yet the weight of the wall and the man are serving to resist this overturning effect. If the man were to jump off the wall, the wall would be more likely to blow over. So, again, the maximum load (dead + live + wind) doesn't give rise to the worst case.

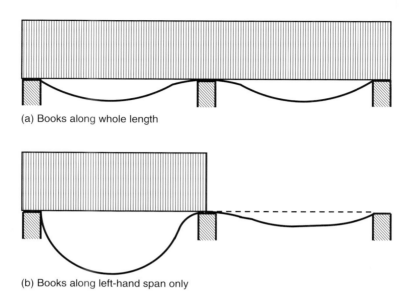

(a) Books along whole length

(b) Books along left-hand span only

Fig. 25.2   Timber bookshelf loaded with books.

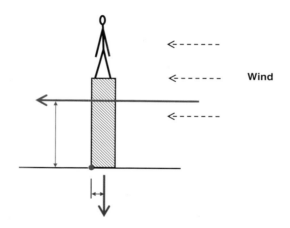

Wind

Fig. 25.3   Man standing on a wall.

For this reason, appropriate load combinations have to be considered in design. You'll learn more about this when you come to study a structural design module as part of your course.

Calculation of maximum shear force and bending moments

If you've read Chapter 16 you will know how to draw shear force and bending moment diagrams. You have to do this before attempting any structural design, as it is important to know the maximum shear force and bending moment in any given situation so you can design against them. Sometimes the case you're dealing with is a standard case and can be dealt with by using some of the standard results discussed in Chapter 16.

Applied shears and moments, and moments of resistance

Normally, structural design entails calculating the moment of resistance (in other words, the maximum bending moment it can sustain without collapsing) of, for example, a beam, then checking that the maximum applied moment (which is the maximum bending moment it is ever actually likely to experience) does not exceed that figure. If, for example, a beam is to be designed to resist a maximum applied bending moment of 400 kN.m, and the standard beam section selected is calculated to have a moment of resistance of 420 kN.m, then all is well.

The same principle applies to shear force: the shear resistance of a beam must exceed the maximum shear force it's ever likely to experience if the design is going to be safe.

Concrete

Concrete is an extremely versatile material: by its nature it can be cast into any shape or form, and therefore it is widely used in construction. Concrete is used for beams, slabs, columns, and foundations. See Fig. 25.4.

Fig. 25.4 **Steel building being constructed around concrete core. The core provides overall stability for the building.**

Reinforcement

As has been mentioned previously, concrete is strong in compression but weak in tension, which severely limits its usefulness in any structural element which experiences bending. Therefore steel bars are placed in the concrete. These steel bars are termed *reinforcing bars*, or *reinforcement*, and being strong in tension, the steel bars help the concrete in catering for tension. The structural design of reinforced concrete elements is largely concerned with determining the number, size, type and location of reinforcing bars, as these factors are critical.

Concrete strength classes

Concrete comes in various compressive strength classes; typical values are 30 or 40 N/mm². Whenever wet concrete is poured on a construction site, samples of the concrete are tested. Slump tests are performed there and then on site: the slump test involves filling a steel cone with concrete under carefully controlled conditions, then removing the cone and measuring the vertical distance by which the concrete cone deforms, or 'slumps'; this gives an indication of the workability of the concrete. Other samples of the concrete are placed into a standard size steel cube mould (steel cylinders are used elsewhere in Europe) left to set and gain strength, and the hardened cubes are tested in a compression testing machine a set period of time after pouring – usually 7 days and 28 days.

Concrete strength classes are based on what is termed the *characteristic* strength of concrete. It is recognised that if a large number of cubes of the same concrete were to be tested, not all samples would fail at exactly the same strength; many would fail at slightly higher or lower values. The characteristic strength is a statistically derived value, and is the value below which not more than 5% of the samples would fail.

For various reasons the same batch of concrete will have different apparent strengths if tested in the form of a cube than it would if tested in the form of a cylinder. The cylinder strength is always less than the cube strength (between about 75% and 85% of the value) and, in Eurocodes, strength classes incorporate both values e.g. strength class C30/37 has a cylinder strength of 30 N/mm² and a cube strength of 37 N/mm².

Reinforcement types

There are two types of reinforcement: mild steel and high yield (or high tensile) steel. As you might infer from the names, high yield steel is stronger than mild steel. Mild steel has a yield strength of 250 N/mm² whereas high yield steel has a yield strength of 460 or 500 N/mm².

Mild steel bars are perfectly round with a smooth surface, whereas high yield steel bars are 'deformed', which means that their surfaces are ribbed. As mentioned above, the steel reinforcement is there to take the tensile forces from the concrete, but it can only do it if there is an efficient means of transferring these forces from the concrete to the steel. A good (but not perfect) analogy is a car tyre. Good tread on a car tyre provides a reasonably high amount of friction between the car and the road, which means the car will stop efficiently when the brakes are applied. By contrast, a bald tyre (no tread at all) will ensure little or no friction between the car and the road, making

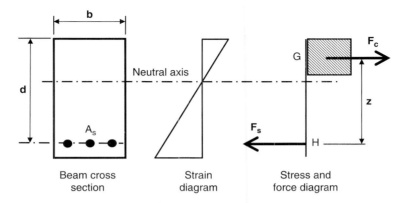

Fig. 25.5   Stresses and forces in reinforced concrete beams.

braking largely ineffective. In the same way, the ribs on the reinforcing bars improve the friction, or 'grip' between the concrete and the steel, meaning that the tensile stresses can be more effectively transferred between the two materials.

Reinforcement bars come in standard sizes. In the UK these are diameters of 6, 8, 10, 12, 16, 20, 25, 32 and 40 mm.

Elastic and plastic behaviour

Most materials, when initially loaded, exhibit what is termed *elastic* behaviour. In this situation, load is proportional to extension, and thus a graph of load v. extension (or stress v. strain) is a straight line. Furthermore elastic behaviour means that if the load is removed it will return to its original shape – this is easily demonstrated by stretching a rubber band then releasing it. (See the early part of Chapter 18 for a reminder of this.)

Beyond a certain point (the *yield point*) the material starts to yield. This is usually characterised by greatly increased extension for a relatively small increase in load. The material is now experiencing *plastic* behaviour. In this situation, load is no longer proportional to extension, and stress is no longer proportional to strain, so the stress v. strain diagram is no longer a straight line. The material also experiences permanent deformation. Eventually, the material will fracture, and failure takes place.

Elastic design, which is design on the basis that the material won't yield under the loads to which it is likely to be subjected, is clearly always going to be safe. However, plastic design, within certain limits, can be justified, on the basis that it leads to a more economic design (because the material is being used more efficiently) and that collapse cannot occur because of the use of factors of safety.

Fig. 25.5 illustrates the principle of plastic design of concrete and the basis on which structural design of beams is carried out. A full description, accompanied by mathematical derivations, can be found in any textbook on reinforced concrete design, but a brief outline is given below. This is just intended to give you a flavour of how it's done, so, if it has already been a particularly long day, feel free to skip the next few paragraphs.

If any reinforced concrete beam is loaded it experiences bending, and hence a bending moment. We need to calculate the bending moment that a given beam can sustain, so we can check that this moment (the moment of resistance) is greater than the maximum moment that will be applied to the beam.

As we discovered earlier in this book, if a beam is sagging it will experience compression in the top and tension in the bottom. In the strain diagram shown in Fig. 25.5, the beam experiences a compressive strain in the top, and tensile strain in the bottom. The level of zero strain, indicated in the figure by the chain-dotted line, is the neutral axis. As we saw in Chapter 19, the neutral axis is half way down the section if the beam is symmetrical and made out of a single material, but in reinforced concrete there are two materials (concrete and steel reinforcement) so the neutral axis will not normally be half way down.

In elastic design the shape of the stress diagram would be the same as the strain diagram. However, we are designing for plastic behaviour, and the shape of the stress diagram in the compression part of the section (that is, the top part) can be approximated to a rectangular block, as shown in Fig. 25.5. As concrete is so poor in tension, we assume that the concrete takes no tensile stress at all, and therefore all the tensile stresses are sustained by the steel reinforcement.

Remember: stress = force/area and therefore force = stress × area.

The force in the concrete, F_c, can be determined by multiplying the stress in the concrete (which is the concrete strength multiplied by various 'adjustment factors') by the area of the concrete (which will be b, the breadth of the section, multiplied by the depth of the stress block).

Similarly, F_s will be the (adjusted) strength of the steel reinforcement multiplied by the cross-sectional area of the reinforcing bars.

Moment about G = $F_s.z$

Moment about H = $F_c.z$

For horizontal equilibrium, $F_s = F_c$ and therefore the two moments must be the same. If we know the maximum moment that will be applied to the beam, we can thus calculate the area of reinforcement required to resist that moment.

Reinforced concrete beams

Reinforced concrete beam design involves assuming a beam size (based on experience and rules of thumb) then using equations from a design standard (which are derived from an approach similar to that shown in Fig. 25.5) in order to determine the area of reinforcement required, in mm². Using standard 'ready-reckoner' bar area tables the designer can determine how many standard-size bars of a particular size are required.

Further formulas can be used to determine the amount of reinforcement (provided in the form of links) to resist shear, and a deflection check can be made. See Fig. 25.6.

- typical reinforced concrete beam size: 200 mm wide × 400 mm deep
- typical reinforcement bar sizes in beams: 16, 20 or 25 mm diameter for main bars, 8, 10 or 12 mm diameter for links

Reinforced concrete slabs

Reinforced concrete slabs are designed on the same basis as beams; they are usually treated as a series of beams, side by side, each of width 1 m. Shear is not normally a problem in conventional slabs, but checks still have to be made. See Fig. 25.7.

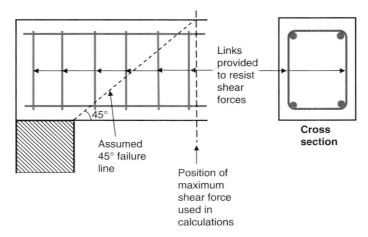

Fig. 25.6   **Typical shear reinforcement in a reinforced concrete beam.**

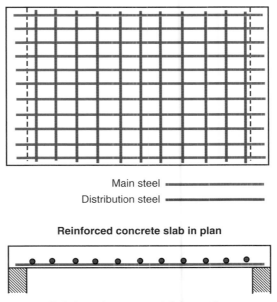

Fig. 25.7   **A reinforced concrete slab.**

- typical reinforced concrete slab depth: 150–250 mm
- typical reinforcement bar sizes in slabs: 10, 12 or 16 mm
- typical reinforcement spacing: 100–200 mm

Reinforced concrete columns

Normally there will be four longitudinal steel bars in a column, one in each corner. The basic design of columns is straightforward but can be complicated by any of the following three possibilities:

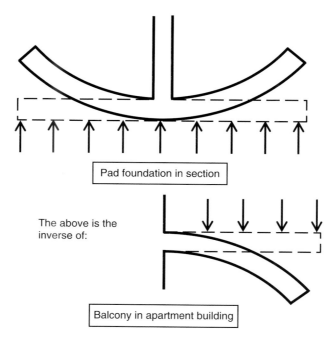

Fig. 25.8   Pad foundations – design principles.

1)   The columns may be slender.
2)   The building may be unbraced, in which case the column is called on to assist in resisting wind loads.
3)   The load in the column may not be purely axial.

The links are nominal and are determined using rules given in design standards.

- typical column cross-sectional size: 300 mm × 300 mm or 400 mm × 400 mm
- typical reinforcement bar sizes in columns: 25, 32 or 40 mm
- typical link size: 12 mm or 16 mm

Reinforced concrete foundations

The various types of concrete foundations – strip, pad, raft and piled – were introduced in Chapter 3. The first three of these are sized on the basis that they spread the load from the superstructure into the ground such that the ground's bearing capacity (that is, the maximum pressure that the ground can sustain) is not exceeded.

Strip foundations are often unreinforced, but if they need to be reinforced then the reinforcement is designed as for pads.

Pad foundations are designed on the basis that the outstand parts of the base will bend as cantilevers under the upward thrust of the soil, as shown in Fig. 25.8a. This can be considered as the same case as the loading on a balcony in an apartment block, turned upside down – see Fig. 25.8b.

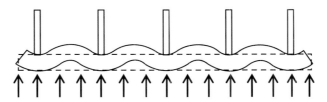

Fig. 25.9   Raft foundation – design principles.

Raft foundations are designed on the same basis as a continuous beam turned upside down. See Fig. 25.9. Broadly speaking, pile caps are designed on a similar basis, but the reader should refer to specialist textbooks for more information on this.

Reinforced concrete detail drawings

As you'll have understood from the foregoing, the reinforcing of concrete is not a question of incorporating some steel in the concrete in a random fashion. In fact, the quantity and exact positioning of the reinforcing bars is crucial, and therefore design and drawing of the reinforcement is an exact science. As far as the drawings are concerned, you'll appreciate that trying to convey, through the medium of drawings, the exact positioning of a complex arrangement of steel bars in space is quite a difficult undertaking. Even if you don't produce the drawings yourself, as an engineer you'll certainly need to understand the conventions used in these drawings, and you'll need to be skilled at interpreting the information contained in them. A very simple example of a reinforced concrete drawing is given in Fig. 25.10.

Along with reinforced concrete drawings, **bar bending schedules** need to be produced. These are tabulated inventories of the number and length of bars of each type required in a particular project or part of the works.

Prestressed concrete

You are no doubt familiar with the action of removing a book from a bookshelf. Library and bookshop staff may need to remove several adjacent books at once if shelves are to be reorganised, or stock introduced or moved in bulk. It is of course possible to remove say half a dozen books individually, but it would be much more time-efficient to move all six at once. This can be done by using both hands to squeeze the six books, thus applying compression to the block of books. This compressive action holds the six books together as one unit while they are moved – see Fig. 25.11. It helps if the six books are of similar size and format. The bigger and heavier the books are, the greater the compressive force that would be required to move them in bulk without the inner ones falling out. Some books (e.g. very heavy encyclopaedia volumes) may be difficult or impossible to move by this means, but a few light paperbacks should prove relatively easy.

Now consider an alternative. Rather than have the librarian's hands apply an external compressive force to the books, imagine that the compressive force is applied by internal means. Moving away from the practicality of the bookshop example, it would be possible to drill a hole through the centres of the six books, and thread a steel cable through them. The steel cable could be stretched tight and secured at its end to two large steel

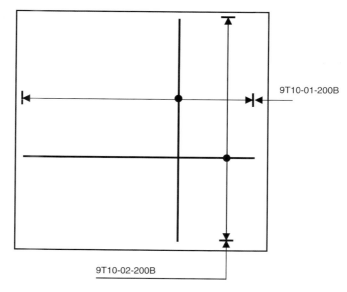

9T10-01-200B

9T10-02-200B

Key: 9 = number of bars in range
T = type of reinforcing steel (high tensile)
10 = bar diameter (10mm)
01 and 02 = bar mark (a unique reference number)
200 = spacing of bars (200mm)
B = bottom face of concrete

Fig. 25.10   Simple reinforced concrete detailing drawing.

Fig. 25.11   Books being moved 'in a block' by hand.

plates laid against the two outer faces of the books. This system would also supply a compressive force to the set of books and keep them together, as shown in Fig. 25.12.

This is a crude representation of a prestressing system. Note the earlier reference to heavy encyclopaedias. The compressive force in the block of books is

Fig. 25.12   **A simple prestressing system involving books.**

incorporating – and, to some extent, opposing – the downward force (due to gravity) from the heavy books.

This concept can now be extended to a concrete beam. As with the books, a steel cable could be threaded through a longitudinal hole in the concrete beam and pulled tight, thus imparting a compressive force to the concrete, allowing the beam to take heavy loads.

Note that although the concrete is subjected to *compressive* forces, the steel cable making this possible is in *tension*. (As an undergraduate student it took me some time to understand this basic concept.) As in any system in equilibrium, the tensile and compressive forces will be equal and opposite, and thus will cancel each other out.

As the series of diagrams in Fig. 25.13 shows, the tensile stress induced in the bottom of a beam can be reduced, or eliminated entirely, by the compressive forces induced by prestressing. For this reason, prestressed beams are 'stronger' than non-prestressed beams, therefore they can be smaller and slenderer in cross section and can span longer distances. As mentioned in Chapter 23, this makes prestressed beams a popular choice in situations where long spans are required or desirable, such as in multi-storey car parks, theatres and other auditoriums, and sports halls.

There are two systems of prestressing: these are **pre-tensioning** and **post-tensioning**. As the names suggest, the distinction between these two lies in the order in which the pouring of concrete and the prestressing are carried out. In pre-tensioning, tensile forces are applied to the cables before the concrete is poured around them. In post-tensioning the concrete is cast around ducts into which steel cables are later inserted; the cables are then tensioned and anchored at their ends by anchor blocks and plates which maintain the prestressing force.

(Note that there is no such thing as 'post-stressing': the word simply has no meaning. If you hear somebody use the word post-stressing, they are possibly referring to post-*tensioning*, which is the more common type of prestressing.)

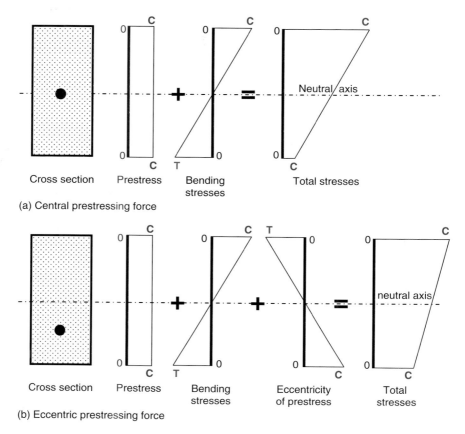

Fig. 25.13 Principles of prestressing.

Precast concrete

The term precast concrete refers not so much to a type of concrete as to the time and place at which the concrete element is constructed – that is, in advance, away from the site.

The most commonly encountered precast concrete elements are slabs, but beams and columns can be precast too. The advantages and disadvantages of precasting were mentioned in Chapter 22.

Masonry

The word 'masonry' means 'that which is crafted by a mason', and traditionally this is stonework. There is nothing new about the use of stone as a building material; it has been used for thousands of years, as manifest in the temples, tombs and other buildings constructed by the ancient Greeks, Egyptians and Romans up to 4000 years ago. (It's sobering to consider how few of the buildings constructed today may still be standing in 4000 years' time.) Many medieval castles and cathedrals were built of stone many centuries ago and are still standing today. Stone therefore is a building material which has stood the test of time.

Fig. 25.14   Stonework in the Royal Crescent, Bath.

Fig. 25.15   Modern housing in brick.

The meaning of the term 'masonry' has now been expanded to incorporate brick-work and blockwork, which have similar characteristics to stonework.

Masonry is a very common material in domestic structures in the UK, thanks largely to its ready availability. Stone is extracted directly from quarries, whereas brick is obtained from clay which is fired in a kiln until it undergoes vitrification – the change of state from soft, mouldable clay to hard, brittle ceramic brick. Certain parts of the UK and elsewhere are recognised for the distinctive type of local stone used in building construction, for example Bath, the Cotswolds, the Lake District, Harrogate, Glasgow, Aberdeen and Paris. Incidentally there are good reasons why stone tends to 'stay local': it is heavy and therefore difficult to transport. See Figs 25.14 and 25.15.

Fig. 25.16   Column spacing.
The Romans realised that stone, being weak in tension, could not span far and hence this surviving Roman building in Nimes, France, is typical of classical architecture in that the supporting columns are closely spaced.

Like concrete, all forms of masonry are strong in compression but weak in tension. But unlike concrete, masonry cannot be reinforced – although there has been research on these lines, it hasn't caught on. Therefore masonry cannot be used in situations where tension can occur, though it can tolerate the small amount of tension that is induced in eccentrically loaded walls or columns. So masonry cannot be used for beams and slabs, because these experience bending, and hence tension, but masonry can be used for compression elements such as walls and columns – see Fig. 25.16. Masonry is also a suitable material for arches, which are in compression throughout.

In this chapter we won't consider further the design or analysis of masonry arches, which are technically complex. Instead, we'll concentrate on the main form of masonry: walls.

In the UK, the external walls of buildings constructed in the past 80 years or so are of cavity wall construction, which comprises two walls, or leaves, separated by a narrow gap. The external leaf is usually of brickwork, around 100 mm thick. The inner leaf is of blockwork, typical thicknesses 100, 140 or 200 mm. The two leaves are separated by a gap of 50 mm (sometimes 75 mm) which may be filled with insulation, though air is a highly effective insulator. The two leaves are connected by steel ties at appropriate intervals horizontally and vertically.

In cavity wall construction it is the inner (blockwork) leaf which takes all the vertical loads (from the building's roof, upper floors, etc), whereas the outer (brick) leaf supports only its own self weight, though it is of course subjected to wind loads as well.

Load-bearing internal walls in buildings are typically of brick or blockwork.

Masonry comprises two main elements: the bricks, blocks or stones themselves, and the mortar which 'glues' them together. There are a number of factors that affect

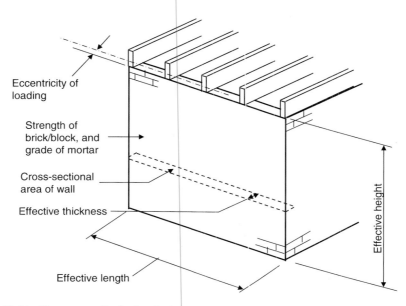

Fig. 25.17   Masonry walls design factors.

the design of masonry. These are illustrated in Fig. 25.17, which depicts the brick back wall of a typical domestic garage, and are listed below.

- *The strength of the individual brick, block or stone.* These strengths vary according to the type of stone or brick.
- *The strength of mortar.* The relevant British Standard lists four mortar designations, along with the composition of each. There is a trade-off between the strength of the mortar and its ability to accommodate movement. The strongest mortar is the least able to accommodate movement, while the weakest of the four is well able to accommodate movement.
- *The thickness of the masonry wall.* When we consider a cavity wall, which has two thicknesses of masonry, design standards specify an effective thickness (i.e. an equivalent thickness of a single wall) for the purposes of calculation.
- *The height of the wall.* This is the vertical distance between what are termed **effective lateral restraints.** An effective lateral restraint stops the wall from moving horizontally: typical examples are floors, roofs and foundations. Design standards also specify an effective height, which is a multiple of the height of the wall dependent on how rigid (or otherwise) the lateral restraint is. For example, timber joists provide a lesser resistance to horizontal movement than a concrete slab would.
- *The width of the wall.* This is the horizontal distance between effective lateral restraints, which in this case are usually walls at right angles to the wall in question. Again, an effective width can be calculated, which varies according to the degree of restraint provided. For example, is the other wall bonded into the wall being considered, or not?
- *The eccentricity of loading.* Does the load that the wall is supporting act down the centreline of the wall, or not? Design standards give guidance on this. Usually a

wall is supporting several loads with different lines of action, so an overall eccentricity (in millimetres) has to be calculated. Eccentricity can lead to bending of the wall: the greater the eccentricity, the greater the bending.

- *The slenderness of the wall.* This is the ratio of the wall's height to its thickness, or more precisely, the ratio of its effective height to its effective thickness. The more 'tall and slim' the wall is, the greater the potential bending.
- *Capacity reduction factor.* This is a function of the eccentricity of loading on the wall and its slenderness, both of which were discussed above. The greater the eccentricity and the greater the slenderness the less load the wall can take, and therefore the smaller the capacity reduction factor. The capacity reduction factor is a figure less than one, which is multiplied by the apparent load-carrying capacity of a wall to give its true load-carrying capacity. For example, a wall whose initial load-carrying capacity is 1000 kN per metre length will in fact only take a load of 240 kN/m if its capacity reduction factor (due to eccentricity and slenderness) is determined as 0.24.

Other factors

The strength of a brick wall is much less than the strength of an individual brick.

As a university lecturer I've many times conducted a laboratory session in which various samples of brick and mortar are tested to destruction in a compression testing apparatus. First of all a solid cube of mortar is tested, of side dimension 70 mm. As expected, it's not very strong, and fails under a compressive stress of 2 N/mm². A single brick is then tested and, as might be expected, it is far stronger than the mortar cube. If engineering brick is used the compressive strength recorded can on occasion exceed 100 N/mm², but a typical value is 80 N/mm². Finally a brick panel is tested in compression. This brick panel comprises three bricks stacked on top of each other, with mortar in the joints; in other words, it is a simple brick wall. Again, results vary a bit, but a typical figure is 30 N/mm² for the compressive strength of a brick wall. Thus we find that the strength of a brick wall (30 N/mm²) is considerably less than the strength of the individual bricks from which it is made (80 N/mm²). This demonstrates the danger of designing a brick wall based on the strength of an individual brick.

Incidentally, brick is a brittle ceramic material and its failure mode is as unpredictable as that of a china plate dropped onto a hard floor surface.

Design standards also incorporate what is often known as gamma (γ) factors, which are factors of safety based on the degree of workmanship and level of supervision of bricklaying operations. As you might imagine, this is somewhat subjective and the default value is usually taken as 3.5.

Design for vertical loads (dead and live)

A critical point in the masonry wall is identified. This is usually low down in the building (for maximum load) and immediately below a floor (to maximise eccentricity). A popular point is immediately below the first floor of the building being designed.

The total vertical load acting at the point concerned is calculated. This may be made up of loads with different lines of action, so a resultant force, at a calculated resultant line of action, is determined.

The eccentricity (that is, the horizontal distance between the resultant force's line of action and the centre line of the wall) is calculated.

The slenderness ratio (that is, effective height divided by effective thickness) is calculated.

Using tables from design standards, the capacity reduction factor is determined, hence the required compressive strength of the wall is calculated.

A combination of brick or block strength and mortar type is chosen whose strength exceeds the required compressive strength.

Piers

Where walls are long and high without frequent lateral supports (for example, the walls of a sports centre) they might be strengthened at regular intervals by brick piers. The strengthening effect varies according to the spacing of the piers, and the plan dimensions of the piers related to those of the wall, and these factors influence the design of the wall, which is otherwise similar to the design procedure outlined above.

Design for lateral loads (e.g. wind)

When it is subjected to lateral loads a masonry wall is designed as a vertical beam or slab which may be one-way spanning (between top and bottom) or two-way spanning (top to bottom and side to side).

Timber

Despite overzealous logging in certain parts of the world, there are still plenty of trees around, which is fortunate, because wood is an excellent building material. Unlike the other materials, which can only be converted into a usable form through sometimes expensive processes, wood can be used in its natural form. Wood that is cut and prepared for constructional use is called *timber* (*lumber* in North America).

A tree trunk is compressed by the weight of the tree above it, and is also resisting short-term gusts of wind on blustery days. A piece of timber therefore has good axial compression properties, hence it can be used for columns, and is also strong in bending, so can be used for beams.

Timber is widely used in the UK in domestic floor construction, where joists – rectangular timber beams whose depth is usually much greater than the breadth – support floorboards, and are typically spaced between 400 and 600 mm apart. Some modern British housing is of timber-framed construction, which is a structure type comprising timber joists supported by closely spaced timber columns. More traditional timber columns can sometimes be seen in older low-rise buildings. Not surprisingly, timber construction is common in parts of the world with lots of trees, for example Scandinavia, Russia and North America. See Figs 25.18 and 25.19 for imaginative uses of timber.

Timber's main problem is that it is not very strong, either in tension or compression, so it can only be used in lightly loaded structures with short spans, such as housing. The exception to this is glued-laminated timber, which is discussed below.

Fig. 25.18   Seeing the wood for the trees.
Tree-like timber columns support the roof of this shopping centre in Maastricht, Holland.

Most timbers used structurally are so-called softwoods. Softwoods come from coniferous trees, that is, the evergreen trees that do not lose their leaves in the winter. Softwoods grow quickly and have very long, straight trunks, which make them ideal for cutting into long beams. Hardwoods can be categorised as either temperate or tropical, depending on the climate of the part of the world in which they are found. Temperate hardwoods come from broad-leafed deciduous trees such as oak and maple tend to be too expensive for most structural use, though they have a better natural resistance to decay than softwoods. Tropical hardwoods are becoming scarce, but because of their natural durability they are used in demanding environments such as external or marine applications.

The moisture content of timber can greatly affect its strength and durability, so pieces of timber are dried out, either naturally or artificially. This drying process is termed *seasoning*. Design standards, which classify timber according to moisture content ranges, are determined by the environment in which the timber finds itself (e.g. heated or unheated buildings).

Fig. 25.19   The ripple effect – a timber gridshell roof
This undulating roof to the Savill Building in Windsor Great Park comprises timber beams forming a square grid structure.

If you examine a piece of timber you will see a grain pattern: a system of lines within the timber which are produced by the living tree's annual growth cycle. Timber is an anisotropic material, which means that its properties vary depending on the orientation of the grain.

Any piece of timber is likely to contain defects, which may be naturally occurring (e.g. knots, waney edge, etc.) or as a result of seasoning (e.g. warping). These defects need to be taken into account in any assessment of the timber's structural integrity. This assessment may be carried out visually by a suitably qualified person, or the timber can be tested by bending in a machine to determine the Young's modulus (E) value, which gives a good indication of the timber's strength.

Because timber is a natural material rather than a manufactured product it is not uniform or consistent in structural quality, and therefore for structural purposes it is grouped into separate categories according to strength. There are a large number of different species of timber, and varying qualities of individual pieces of timber (because of the presence of defects as discussed above), so structural design standards categorise these into strength classes. In British Standards and Eurocodes, the strength classes are designated as either C (for coniferous trees) or D (for deciduous trees) followed by a number which represents the timber's bending strength. For example, the most common strength classes, C16 and C24, represent coniferous trees of bending strength 16 N/mm^2 and 24 N/mm^2 respectively.

Broadly speaking, the British Standard technique involves calculating an actual stress from structural mechanics principles, then calculating a permissible stress and comparing the two. Provided the permissible stress is greater than the actual stress, then the design is safe. Permissible stresses are obtained by multiplying a grade (or 'basic') stress by so-called *K factors*.

Design standards refer to a large number of K factors, but only a small number are required for most calculations. These K factors relate to:

- *Duration of loading.* This recognises that a timber roof may be subjected to snow loads which may linger for days or even weeks, whereas a timber floor within a building will not experience such loads.
- *Depth of joist.* Design standards assume a given depth of timber joist (e.g. 300 mm), so a conversion factor is required for other depths.
- *Load sharing.* If you are standing within a bedroom on the first floor of your house – which in the UK is likely to comprise timber joist supporting timber floorboards – it is unlikely that the joist immediately below your feet will carry all your load. In practice, the load will be shared between a number of joists. So design standards assume that if the spacing of joists is below a certain figure (e.g. 610 mm) then load sharing takes place and a load sharing factor (usually 1.1) is used in calculations. If the joists are spaced further apart than this, there is no load sharing and the load sharing factor would be 1. Load sharing can be regarded as an 'on-off switch' therefore; it either occurs or it doesn't.

Normally the size of a timber beam (or column) is limited by the size of the tree trunk from which it is cut, but there is an exception to this. Laminates – narrow strips of timber of typical size 100 mm wide × 40 mm deep – can be glued to each other to produce beams of considerable depth – one metre or more – and their ends can be glued together to produce long-span timber beams. This type of composite timber beam is called a glued-laminated timber (glulam for short).

Structural elements in timber

Timber can be used to form beams (floor joists), columns and trusses used for roof systems.

Glued-laminated (glulam) products include beams or slabs, the latter being in the form of cross-laminated flooring systems (architects love these!) which can also be used to form timber walls. As mentioned above, glulam beams achieve greater spans and can carry greater loads, but are expensive, so are rarely used in the UK. Because timber is chemically inert, glulam beams are a good choice to span swimming pools, as they are not attacked by the chlorine in these environments. Also using glue are ply-web and ply-box beams, discussed more fully below.

Timber connections are complicated, expensive and beyond the scope of this book.

Timber joists

Actual stresses are calculated for each of bending shear, bearing, deflection and lateral stability, and these are compared with permissible stresses in each case. Provided the permissible stress is greater than the actual stress, the design is acceptable.

- typical timber joist size: 50 mm wide × 200 mm deep
- typical timber joist spacing: 400–600 mm

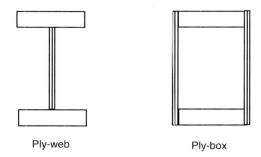

Ply-web   Ply-box

Fig. 25.20   **Ply-web and ply-box beams.**

Timber columns

As with other materials, slenderness is an important factor, as is the effective length of the column. Actual compressive stress (= force/area) is compared with a permissible stress derived from K factors.

Glulam beams

The design procedure is similar to that for timber joists, though there is an extra K factor to take into account the number of laminations (individual layers of timber) in the beam.

Ply-web and ply-box beams

These both incorporate solid timber and plywood, glued together. In each case, the plywood forms the vertical web(s) and the timber forms the horizontal flanges. The ply-web is an I section whereas the ply-box is a box section. They are rarely used where appearance is important because squeezed glue lines are apparent and finishing involves extra cost. See Fig. 25.20.

Calculations for these beam types centre on demonstrating that the permissible tensile and compressive stresses in both the timber flanges and the plywood webs are greater than the actual maximum stress in the beam.

Steel

Steel is an extremely useful structural material and plays a part in most buildings that are larger than domestic in scale. Steel is strong in both tension and compression, making it useful for high-rise buildings and low-rise framed structures, single-storey (large) sheds and long-span roof structures. Buildings whose main structure is of another material (such as concrete or masonry) often have pitched roof structures made out of steel. For long-span structures such as suspension bridges or buildings where long spans are required (such as sports stadia) steel is often the only viable option. See Figs 25.21 to 25.24 for examples.

Steel is vulnerable to fire and corrosion unless appropriate protection is provided and maintained, but the main problem with steel from a designer's point of view is the possibility of buckling.

Fig. 25.21   **St Pancras station, London.**

Fig. 25.22   **Leeds railway station as rebuilt in 2002.**

Unlike concrete, steel is, by its very nature, prefabricated, and arrives on construction sites as a finished product, ready to be erected as part of a structure. Steelwork is manufactured in the form of standard types of steel component, as follows:

- *Universal beam (UB)*. This is shaped like the letter I in cross section, and, as the name suggests, is usually used for beams.
- *Universal column (UC)*. This is shaped like the letter H in cross section and is used for columns (or stanchions, as steel columns are sometimes called) and sometimes for beams as well, particularly if restricted headroom is a problem.

Fig. 25.23   Making the connection.
A substantial steelwork connection at Heathrow Airport, Terminal 5.

Fig. 25.24   Cable-stayed bridge, Dublin.

- *Angles (equal or unequal).* These sections are shaped like the letter L in cross section; the horizontal and vertical parts of the L might be equal in length (equal angles) or of differing lengths (unequal angles).
- *Tees.* These sections are shaped like the letter T in cross section, and are usually created by cutting an I section in half longitudinally at its mid depth.
- *Circular hollow sections (CHS).* Hollow sections are essentially tubes. They are favoured by architects as, when used with welded connections, very visually clean lines can be achieved. They are often used as the components of lattice girders or space frames in roof structures.

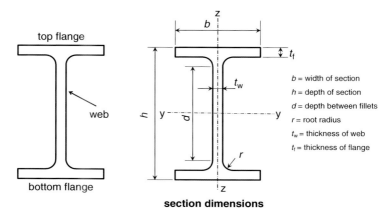

top flange

web

bottom flange

section dimensions

b = width of section
h = depth of section
d = depth between fillets
r = root radius
t_w = thickness of web
t_f = thickness of flange

Fig. 25.25 A typical steel section.

- *Square hollow sections (SHS).* These are tubes which are square in cross section.
- *Rectangular hollow sections (RHS).* These tubes are rectangular in cross section.

The vertical part of a steel cross section is called the **web** and the horizontal part (or 'crossbar') is called the **flange**, as illustrated in Fig. 25.25.

Each of these types of steel component comes in standard sizes. Steel manufacture in the UK is carried out by Tata (formerly Corus, and before that British Steel) who publish tables of properties of each standard section for use in design.

Designation of steel sections

Universal beams are designated in the form $A \times B \times C$, where:
  A is the nominal depth of the cross section
  B is the nominal breadth of the cross section
  C is the mass of the section in kg/m.
  For example, a $203 \times 133 \times 25$ beam has an approximate depth of 203 mm, a breadth of approximately 133 mm and a mass (sometimes called 'weight') of 25 kg/m.

Steel connections

The structural steelwork components listed above have to be joined together to form a structure. These joints are termed **connections**. The design of connections is quite an art (and a science) in itself, and the fabrication costs of complex connections can be considerable.

There are essentially three types of connection: riveted, bolted and welded.

- **Riveted** connections involve the use of rivets, a type of steel connector that is now obsolete. Rivets are circular steel rods which are forced into pre-formed holes under high pressure, appropriate heads being formed on the rivets as part of this process. Despite their obsolescence, there are many existing structures that have riveted connections – for example, many steel railway bridges in the UK – and a knowledge of riveted structures is essential for any engineer involved in the maintenance, appraisal or alteration of such structures.

- *Bolted* connections involve the use of nuts and bolts to connect two or more pieces of metal together. Such connections can easily be made on site.
- *Welded* connections involve the use of welds, which join two steel surfaces by fusion or pressure, or both. If done correctly, such connections are very strong indeed, and give complete continuity between the two pieces of metal that have been welded together. Welding is a precision process which is more easily done in a factory or workshop environment. Welding on site is best avoided unless absolutely necessary, and can only be carried out by welders who are qualified (and certificated) in the type of welding being carried out.

Some common connection types are shown in Fig. 25.26.

Buckling

As mentioned above, the avoidance of buckling is an important factor in the design of structural steelwork. We've previously seen that the slenderness of a structural member (for example, a column) is the ratio of the member's length to its thickness; the greater that ratio, the greater the slenderness. The greater the slenderness of a steel member, the more likely it is to buckle.

> *Take, for example, an everyday object that you probably have on your desk at present: a plastic ruler. If you grip the two ends of the ruler and try to pull it apart (that is, apply tension) you will not break the ruler unless you have superhuman strength. Plastic is quite strong in tension. However, if you push the ends of the ruler towards each other, you are applying compression, and it's a different story here. You only need to apply a small compressive force and the ruler will bend, or buckle, because of its slenderness. This propensity to buckle under small compressive forces limits a plastic ruler's usefulness is in compression. See Fig. 25.27.*
>
> *This is not to say, however, that plastic is weak in compression. You cannot realistically expect to crush a plastic ruler with your bare hands. If there were physical restrictions that prevented the ruler from bending at all then the ruler would not buckle, and, in theory, the true compressive strength of the plastic ruler would be realised.*

There are two types of buckling of structural steelwork members:

Local buckling

You are aware that slender columns (in any material) are prone to buckling, and the more slender the column the more likely it is to buckle. Steelwork columns tend to be more slender than those in other materials, hence the possibility of buckling is an important design consideration with steelwork. However, the possibility of buckling is not restricted to columns; the webs and flanges of individual structural sections are likely to buckle if slender.

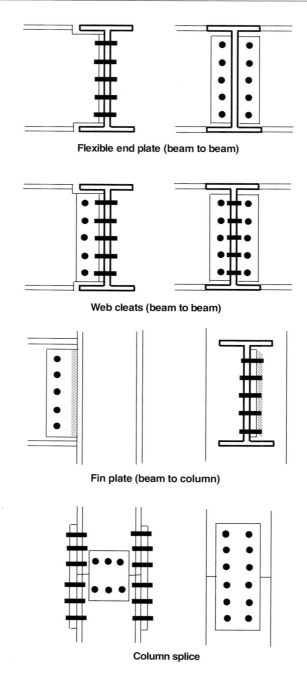

Fig. 25.26 Steelwork connection types.

With columns we have seen that a column's slenderness is the ratio of its height to its thickness. From Fig. 25.28 you will see that the corresponding ratios for flanges and webs are c/t_f and d/t_w respectively. If these ratios exceed a certain amount it is possible that local buckling of the flange and/or the web might occur.

Fig. 25.27   A plastic ruler in compression.

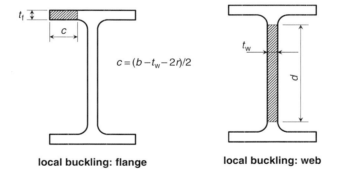

local buckling: flange   **local buckling: web**

Fig. 25.28   Dimensions for local buckling.

If an outstand flange is too slender (that is, if the length of the outstand part is too long compared with the flange thickness) then local buckling of the flange can occur, as shown in Fig. 25.29.

If a web is too slender (that is, if the distance between fillets is too long compared with the web thickness), then local buckling of the web can occur, as shown in Fig. 25.30.

Based on the possibility (or otherwise) of local buckling occurring, standard steel sections are divided into four distinct classifications (see below).

Lateral torsional buckling (LTB)

If a steel section is loaded with a central point load, a vertical deflection will occur, as might be expected. If the section is too slender, then a horizontal deflection will also occur, accompanied by a rotation, or twisting, of the cross section. This combination

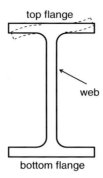

Fig. 25.29   Flange local buckling.

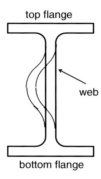

Fig. 25.30   Web local buckling.

of sideways movement and turning is known as ***lateral torsional buckling***, sometimes known by its acronym LTB. See Fig. 25.31.

> *You can demonstrate lateral torsional buckling for yourself if you reach for that plastic ruler again. Grip the ruler firmly at one end; the length of the ruler should be horizontal but orientated so that the flat part of the ruler is vertical. Using a finger, apply a point load to the free (i.e. unsupported) end of the ruler. You will notice that the ruler will move downwards, but also sidewards, and there will be some rotation.*

Clearly, lateral torsional buckling is undesirable. We will talk further about how it can be controlled or eliminated.

The behaviour of steel in tension

If a mild steel bar is subjected to a tensile test in a laboratory, as load is applied some stretching, or extension, of the sample would take place. The amount of this extension is modest at first, and is proportional to the load applied; in other words, the load/extension graph would initially be a straight line. In this initial phase, the sample is exhibiting elastic behaviour and is said to be in the ***elastic zone***. Eventually, a point

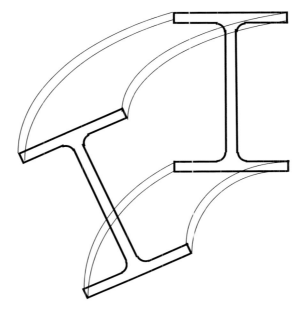

Fig. 25.31   Lateral torsional buckling.

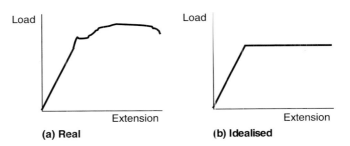

(a) Real   **(b) Idealised**

Fig. 25.32   Load/extension graph for steel.

is reached where the sample yields: its behaviour is no longer elastic, and much larger extensions are achieved with only small increases in load. The sample is now in the *plastic zone*, and the extensions are now permanent and irreversible. Eventually the sample will fail, or break, with a very loud bang. It has been found experimentally that the sample has extended by about 33% of its original length on failure. Fig. 25.32 shows the load/extension graph a) in reality, and b) a simplified ideal version. If you've read Chapter 18 about stress and strain, you'll realise that load can be converted to stress by dividing the load figures by the cross-sectional area of the sample, and extension can be converted to strain by dividing by the original length of the sample. Fig. 25.33 shows the corresponding stress/strain graphs, which are the same shape as the corresponding load/extension graphs.

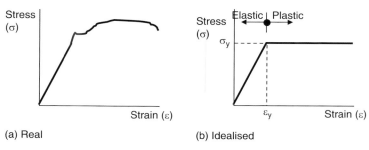

Fig. 25.33   **Stress/strain graph for steel.**

Introduction to plastic design

Most materials exhibit elastic behaviour up to a point. This means that stress is proportional to strain in this elastic zone, and a graph of stress versus strain shows a sloping straight line. This elastic behaviour stops once the stress and strain reach the limit of proportionality, whereupon the material yields and continues to strain without any corresponding increase in stress. Once the material is strained beyond the limiting strain, ε_y, it will exhibit plastic behaviour with a constant stress, σ_y, as the strains increase. See Fig. 25.33 again.

When part or all of a beam or other structural element has been strained beyond the yield point, the structure will start to collapse, and the stress distribution diagrams are shown in Fig. 25.34.

When the beam is in the elastic range of loading, there will be a uniform variation of stress throughout the depth of the beam, as shown in the top diagram in Figure 25.34. As the maximum stresses occur in the top and bottom fibres of the beam under bending (max tensile stress in the bottom, and max compressive stress in the top) it will be these fibres that yield first as the stress on the beam is increased beyond the yield point.

As the load on the beam is increased further the strains across the section will increase and the plastic zone will spread inwards from both top and bottom, as shown in Fig. 25.34b. Note that the stress cannot exceed the yield stress σ_y.

If the load is increased still further, the plastic zone will expand until the entire section is plastic, as shown in Fig. 25.34c. The beam has now reached its full moment capacity, a 'plastic hinge' is formed and the beam will be on the verge of collapse.

Note: Because factors of safety are used in calculations, the use of plastic design should not lead to actual collapse of a designed member in use.

Strength grades

There are currently four strength classes for structural steel: grades S235, S275, S355 and S450. The figure represents the yield strength in N/mm^2, and the S stands for structural (not steel). The most commonly used grade in practice is S275, which is equivalent to the grade historically known as 'Grade 43'.

The Young's modulus (E) of steel should be taken as $210\,kN/mm^2$.

The steel design process is based on the bending equation ($\sigma = M/z$) which we previously encountered in Chapter 19. As stated there, this equation is the basis of all structural design.

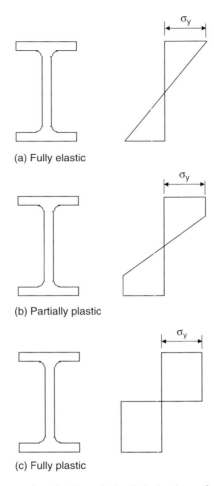

(a) Fully elastic

(b) Partially plastic

(c) Fully plastic

Fig. 25.34   Stress diagrams for elastic and plastic behaviour of steel sections.

Selection of a standard section to try

An adaptation of the above equation is:

Plastic modulus = Max bending moment/steel strength

As we have specified a steel strength, and have calculated the maximum bending moment, we can use this equation to calculate the minimum plastic modulus required in mm^3 units. Divide that figure by 1000 to present the plastic modulus in cm^3.

Using standard tables of steel section properties (produced by Tata, available free from them, and also reproduced in many structural design textbooks) we can select a standard universal beam section whose plastic modulus is equal to, or slightly greater than, the minimum required, and we can progress our calculations using that standard section.

Classification

Sections are classified according to their slenderness and hence their susceptibility to local buckling. There are four classifications, as follows.

Class 1 ('plastic'): The elements are so proportioned that there is no tendency for local buckling and therefore the section can develop its full plastic moment. Also, the plastic section has sufficient rotational capacity to develop plastic hinge action. Design using plastic theory is therefore possible.

Class 2 ('compact'): As with Class 1 sections, the section can develop its full plastic moment, but the rotational capacity is insufficient for the use of plastic theory.

Class 3 ('semi-compact'): The section is capable of reaching the yield stress in its outermost fibres, but can only sustain an elastic stress distribution before local buckling can take place.

Class 4 ('slender'): The tendency towards local buckling is such that a reduced design stress is necessary.

A Class 1 (plastic) section is the most desirable, because no local buckling takes place and the full strength of the section is realised. By contrast, a Class 4 (slender) section is the least desirable as, like our plastic ruler, it will fail in buckling after only a small amount of load is applied and therefore its full potential is not realised.

Restrained beams

As discussed above, buckling of steel beams is to be avoided. Local buckling will not occur if we pick a Class 1 (plastic) section. Lateral torsional buckling will not happen if a beam is physically restrained from moving sideways.

If the steel beam is directly supporting a concrete slab along its entire length – a common situation in practice – friction between the top of the steel beam and underside of the concrete will prevent the beam from moving sideways, and therefore lateral torsional buckling cannot take place. The beam is said to be fully laterally restrained. See Fig. 25.35.

No such restraint is provided if the steel beam does not directly support a slab. However, in practice a steel beam is normally restrained at its ends by supporting columns, and may be stopped from moving sideways at certain points along its length by the presence of secondary steel beams, which are normally orientated at right angles to the main (primary) beam – see Fig. 25.36. Such beams are partially laterally restrained, but it should be noted that they are unrestrained between such points of restraint, and this lack of restraint must be considered in design.

Design of steel beams (if fully restrained)

We need to distinguish between restrained beams (that is, those that are physically prevented from moving sideways) and unrestrained beams. The text below gives an introduction to the design of restrained beams. The design of unrestrained beams is beyond the scope of this book.

Analyse the loads and determine the maximum shear force and bending moments, and the maximum deflection.

1) Select a standard section to try (as discussed above).
2) Determine the section classification (Class 1, 2, 3 or 4) of the chosen section.

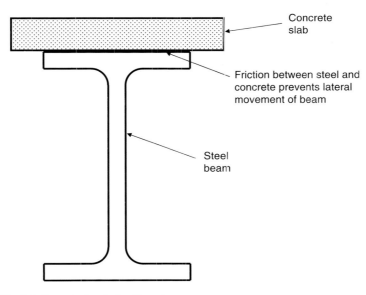

Fig. 25.35   A fully restrained steel beam.

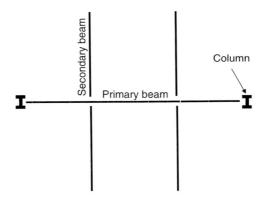

Fig. 25.36   A partially restrained steel beam (in plan).

3) Calculate the moment of resistance of the section and check that this is greater than the maximum applied bending moment.
4) Calculate the shear resistance of the section and check that this is greater than the maximum applied shear force.
5) Calculate the maximum permitted deflection (usually a multiple of the span) and check that this is greater than the actual maximum deflection.

Design of steel columns (if axially loaded)

1) Calculate ultimate axial load applied to the column.
2) Determine the effective length of the column.
3) Select a trial section.

Fig. 25.37   Brighton West Pier.
Closed in the 1970s, this pier has been left to rot since; the photo shows what was left of it in 2009.

4) Calculate the slenderness ratio (= effective length/radius of gyration).
5) Obtain the compression strength of the column from design standards.
6) Calculate the compression resistance (force) of the column from the expression (= cross-sectional area × strength).
7) Finally, check that the compression resistance force is equal to or greater than the ultimate axial load.

Most importantly, every building and structure needs to be maintained throughout its life. Fig. 25.37 shows what might happen if it isn't.

26 More on structural types and forms

In this chapter we'll take a quick tour through the various structures types and forms that have been designed and used over the years – and in some cases, over the centuries. This is aimed to inspire you; I hope it will. The more conventional pieces of structure that are found in most 'ordinary' buildings – beams of various sorts, slabs, foundations, etc – have been discussed in Chapters 3 and 23. We're now moving on to some more exotic types. In each case I'll mention some famous examples of the form, but that is not to say that other, lesser know examples don't exist – maybe even in your home town. I've said it before: look around you at the various buildings you encounter in your travels and in everyday life.

Before reading this chapter you need to be familiar with the concepts of tension, compression, bending and shear, as we'll be using them to explain how each structure type works. Reread Chapter 3 if you're in any doubt.

Some structures are not what they appear to be. For example, London's Millennium Dome is not, technically, a dome at all – it is a cable-stayed membrane structure. Similarly, some 'arches' are not actually arches – see below. And there is also some blurring of the boundaries between vaults, domes and shells.

The types of structure we are going to study in this chapter are:

- folded plate structures
- arches
- vaults
- domes
- shells
- cable structures
- inflatable structures
- space structures
- bridges
- dams
- tunnels

Note that this chapter is simply an introduction to these structure types. Ideas for further reading are given in the bibliography at the end of this book.

Basic Structures, Second Edition. Philip Garrison.
© 2011 John Wiley & Sons, Ltd. Published 2011 by John Wiley & Sons, Ltd.

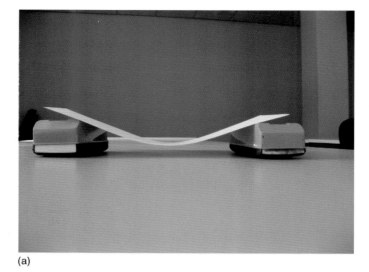

(a)

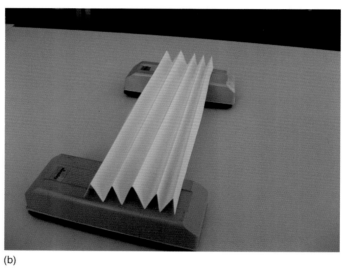

(b)

Fig. 26.1   **(a) Sheet of paper spanning between supports (b) Principle of a folded plate structure.**

Folded plate structures

Take a piece of A4 paper and try to make it span (lengthways) between two supports about 250 mm apart, and you'll find that the paper will simply sag and fall between the two supports, as shown in Fig. 26.1a. Now take the sheet of paper and fold it in a concertina fashion, the series of folds being about 25 mm apart, as shown in Fig. 26.1b, and, again, get it to span between two supports. This time you will find that the (folded) paper will span the 250 mm distance without difficulty, and will even support a light loading such as a pen or two – though anything heavier such as a pair of scissors or a set of keys will make it distort and collapse.

Fig. 26.2   A folded plate roof.
Green Sports Hall, Leeds Metropolitan University.

The folded piece of paper could be made stronger by gluing the ends of the sheet to a rigid vertical end plate.

This principle has been used to make structures which have little rigidity (such as our sheet of A4 paper) into something more rigid and therefore structurally useful. One common example is corrugated iron or steel. Steel sheet is not particularly strong, but its strength is improved by forming it into a sine wave profile and using it for roofing or fencing. A further – more modern – example is the profiled steel decking onto which concrete is poured to form floor slabs.

On a larger scale, folded concrete roofs are a feature of 1960s architecture in the UK; they were used to form long-span roofs over supermarkets and market halls, and can also be found at some dated motorway service areas. An example is shown in Fig. 26.2.

Some structures have been built with more complex folding systems: think large scale origami!

Arches

We are all familiar with an arch as a particular shape, but in structural terms the word 'arch' refers to a particular type of structural behaviour. So sometimes what appears to be an arch is not, strictly speaking, an arch at all.

Consider a wire or cable strung horizontally between two fixed points. If a tight-rope walker were to walk along the wire, the profile of the wire would appear as shown in Fig. 26.3a and b, dependent on the tightrope artist's position (represented by an arrow) along the length of the wire. Now consider a washing line full of newly washed clothes put out to dry: the profile of the loaded washing line might appear as shown in Fig. 26.3c. As you would expect, the wire (or washing line) is in tension throughout its length.

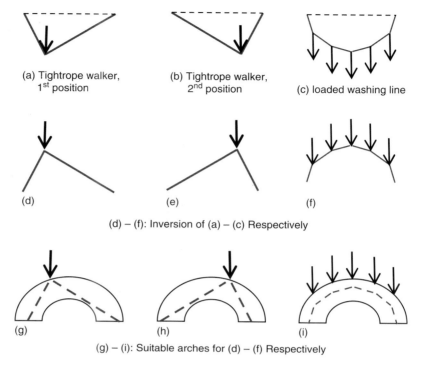

(a) Tightrope walker, 1st position

(b) Tightrope walker, 2nd position

(c) loaded washing line

(d)

(e)

(f)

(d) – (f): Inversion of (a) – (c) Respectively

(g)

(h)

(i)

(g) – (i): Suitable arches for (d) – (f) Respectively

Fig. 26.3 **Cables and arches.**

Now let's reflect each of these diagrams in the horizontal axis which passes through the two ends of the wire, so that the diagram now appears above the horizontal line as a mirror image of what was previously below the line. The resulting profiles will be as shown in Figs. 26.3d–f. Assuming the 'wire' is now made of something more substantial that has compressive strength, each of these profiles now represents an arch.

More precisely, each of these profiles represents the line of action of the forces in an arch that would result from the loading shown. These forces are all in compression, and this is one of the main features of an arch: *an arch is in compression throughout.*

The physical arches that can sustain the forces shown in Figs. 26.3d–f are shown in Figs. 26.3g–i. Note that the line of action of the forces is within the zone of the arch throughout. This is absolutely essential. If the internal forces pass outside the arch profile anywhere, the arch will collapse.

Now take a sheet of A4 paper and, using both hands, form it into an arch shape as shown in Fig. 26.4. The instant you remove your hands the paper arch will collapse. This is because your hands, before you removed them, were providing a horizontal restraint and, the moment that restraint was removed, the piece of paper was free to move horizontally and thus collapse.

This is true of all arches. Any arch has to have some sort of restraint at its two supporting ends (the **springings**) to stop it from deforming and collapsing. This restraint

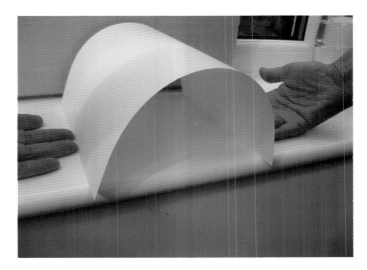

Fig. 26.4   **Piece of paper as an arch.**

Fig. 26.5   **New Wembley stadium under construction.**

might take the form of a natural rock abutment in the case of an arch bridge, or some type of massive buttress. In the case of the new Wembley Stadium, shown under construction in Fig. 26.5, the restraint is provided by the very cables it is supporting. Some bridge arches use the bridge deck as a tension member that ties the two ends of the arch together and thus stops them moving outwards: an example of this is shown in Fig. 26.6.

Arches of various materials are widely used in bridge construction (see below for more on this). They are also sometimes used in buildings, an important example being medieval cathedrals, where massive buttresses are often used to resist the horizontal outward forces at the base of the arch. The angled stone elements shown in Fig. 26.7 – a photograph of York Minster – are so-called flying buttresses. Although they appear to be ornamental, their purpose is purely functional in resisting the

Fig. 26.6   **Tyne Bridge, Newcastle.**

Fig. 26.7   **York Minster.**

outward forces from the roof of the main part of the cathedral, which is at the right hand side of the photograph. The pinnacles shown on the left hand side of the photograph have the function of providing a downward force by virtue of their self weight to maintain the structure in equilibrium. If the pinnacles were to be removed the whole structure might collapse.

So, to summarise, some of the main characteristics of an arch are that:

- lines of thrust generated by the forces on the arch must be physically contained within the arch structure itself, otherwise it will collapse

Fig. 26.8   **St Paul's Cathedral, London**

- an arch is in compression throughout
- arches must be designed to resist the horizontal thrusts at the two ends of the arch.

Vaults

The word 'vault' conjures up images of burial chambers or the strongrooms under banks where valuables might be stored, but in structural terms it simply relates to an arch which is very wide. There could be a series of these forming a roof structure. If they are hemispherical they are often known as barrel vaults. As with arches, vaulted structures need to be buttressed to cater for the horizontal forces. Adaptations of the conventional vault include the cross vault, which is supported at its four corners only, and the pointed vault.

Domes

A dome is 'hemispherical' in shape. Why the inverted commas? Well, most domes are not true hemispheres, geometrically speaking. Some are flattened hemispheres, some are deep hemispheres, and some are not hemispheres at all: they are spherical, or near spherical in shape.

Let us start by considering the near-hemispherical domes found on historic buildings used for religious or government purposes. Examples include cathedrals such as St Paul's Cathedral (see Fig. 26.8) in London and St Peter's Basilica in Rome, the Pantheon in Rome, the Hagia Sophia Mosque in Istanbul, Marmorkirken in Copenhagen (see Fig. 26.9), the Capitol building in Washington DC, along with most US state capital government buildings. Although domes can span a large circular

Fig. 26.9   Marmorkirken, Copenhagen.

space, they are not an efficient way of doing so and are generally present to form an impressive central feature of a grand piece of architecture.

Structurally, domes can be thought of as three-dimensional arches, but the analogy has its limitations. As we have seen, an arch is in compression throughout, whereas tension can – and usually does – occur in a dome, and this tension is the cause of many of the structural problems associated with domes. In constructional terms, an arch cannot function until it is complete, so sometimes substantial temporary works are required for an arch's construction. By contrast, a partially constructed dome can be stable.

Consider half of an orange as representing a dome, as shown in Fig. 26.10a. The lower parts of the orange will bulge outwards as it is loaded. If cuts are made in the base of the hemispherical orange, these cuts will tear upwards under load, as shown in Figure 26.10b. This demonstrates that, whilst the stresses along the lines that radiate from the apex of the dome down to its base (let's call these the ribs) are in compression – as with an arch – the stresses along the circumferential lines around the dome (the hoops) are compressive in the upper part, but tensile in the lower part.

These tensile forces can lead to radial cracking in real domes, though this cracking can be limited or eliminated altogether if the dome is restrained from lateral movement at its base.

The walls supporting a dome may be circular in plan to match the footprint of the dome. However, more usually they are square or octagonal in plan, therefore the dome cannot be seated directly on them, and some form of transfer structure (called a pendentive) is required. See Figure 26.11.

Spherical domes have been used in modern structures such as the domes used at the Expo '68 world fair in Montreal and the Epcot Centre in Florida and, more recently, the dome at the Victoria Shopping Centre in Belfast, Northern Ireland. The new western concourse of King's Cross Station in London (under construction at the time of writing) will be a half dome.

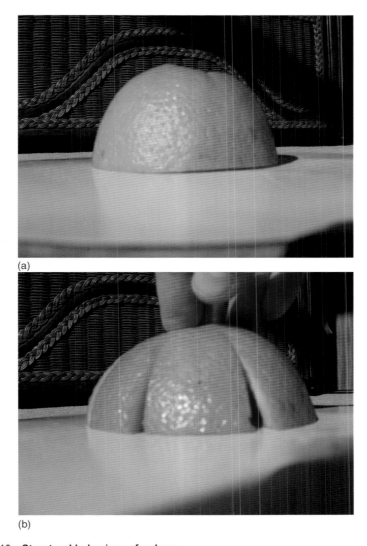

(a)

(b)

Fig. 26.10 Structural behaviour of a dome.

Such domes comprise a pattern of steel or plastic ribs supporting panels of glass or transparent plastic. The ribs are arranged in a repeating pattern of geometrical shapes, the triangle, diamond and hexagon being the most common. Diamatic, lamella and geodesic domes are types of dome with differing rib patterns. Fig. 26.12 shows a modern dome.

Shells

Shell structures resemble arches curved in two dimensions, and as such are often called vaults or domes, depending on their shape. The most famous example is Australia's Sydney Opera House. They are usually made of reinforced concrete cast onto appropriately shaped formwork.

Fig. 26.11   Interior of St Paul's Cathedral, Mdina, Malta.

Fig. 26.12   Cabot Circus shopping centre, Bristol.

A variation on this is monocoque or stress-skinned structures. The structure of a car, ship or plane comprises a framework sandwiched between thin, curved external panels. Occasionally this type of structure finds its way into civil engineering: a modern example is the Media Stand at Lord's Cricket Ground in London, shown in Fig. 26.13.

Fig. 26.13   Lord's Cricket Ground media centre.

Cable structures

Have you ever been camping? If so, and you erected (or helped to erect) the tent, you will have some knowledge of a tent's structure. Tents come in various types and sizes, but a simple ridge tent (for example) comprises two main poles that support the fabric of the tent by holding it at a fixed height above the ground. These poles are in compression. To make the tent stable, guy ropes are attached to the tops of the poles and are secured to the ground using tent pegs at various points outside the tent's envelope. These ropes are in tension.

Essentially, cable structures are like giant tents; they comprise cables in tension supporting a building or bridge structure from above. The cables, in turn, are supported by vertical or inclined masts. The Millennium Dome (now the O2 Arena) in North Greenwich, London, is not a dome at all but a cable net and membrane structure. See Figs 26.14 and 26.15.

Inflatable structures

You will have come across inflatable structures since childhood. Examples include airbeds (inflatable beds used for sleeping in tents, or for floating on water), paddling pools, or various inflatable objects that can be sat on or in, sometimes towed behind a speedboat and used in watersports. Happy days!

It is possible to make an entire structure out of inflatable elements, an example being the bouncy castles found at children's playgrounds. Alternatively, the whole structure can itself be an inflatable element.

Consider a balloon. The balloon can be inflated by blowing into the mouthpiece. When the balloon is full of air, and the mouthpiece tied and sealed, the air inside the balloon is in compression. As you know, for equilibrium, that compressive force has

Fig. 26.14 O2 Arena, London (cable net/membrane).
Despite its shape and original name of Millennium Dome, this structure is not technically a dome at all; it is effectively a giant tent – the roof membrane is suspended by cables from inclined masts.

Fig. 26.15 Wembley Park station (cable stayed).
To provide a largely column-free concourse, the roof of this railway station extension is supported by cables suspended from a mast.

to be counterbalanced by a tensile force, and that tensile force exists in the rubber skin of the balloon. You will know from experience that if you try to overinflate the balloon then the rubber will be stretched beyond its tensile strength and will burst – usually with a loud bang. If the balloon is filled with a gas such as helium, which is lighter than air, it will float off into the sky unless secured to the ground or a solid object – and the securing string will be in tension.

Fig. 26.16   **Concourse roof, Gare Du Sud, Nice, France.**

Inflatable structures are usually temporary structures used for exhibitions or seminars. Their great advantage over conventional temporary structures is the ease with which they can be transported, assembled and dismantled. The structure needs to be kept inflated by a continuous supply of forced air, and double entrances to the structure need to function as air locks to prevent (or at least limit) the exit of the compressed air from the structure. Furthermore, the structure needs to be held down by heavy weights or by guy ropes securing the structure to the ground to avoid the building moving in high winds.

Pneumatic roofs have also been used for larger structures such as sports stadia.

Space structures

These are three-dimensional lattice trusses designed to span large distances in two directions. They often support glazing and therefore can let a lot of light into a building from above. See Fig. 26.16.

Bridges

Bridges really require a whole book on their own (see the Further Reading section for some good books on bridges), but there are three principal types:

1) beam
2) arch
3) suspension

Beam bridges comprise beams spanning on to vertical supports (or piers). The beams may be of timber, steel or concrete, and may be a solid beam or an open web type beam such as a lattice truss. See Fig. 26.17. The supporting piers are usually of concrete

Fig. 26.17   **A59 bridge, Bolton Abbey, North Yorkshire.**

Fig. 26.18   **A traditional arch bridge.**

or masonry. A variation is the cantilever type bridge, in which the bridge deck is built outwards from central supports. The Forth Rail Bridge in Scotland is a famous example of a steel cantilever bridge.

In an ***arch bridge*** an arch, or series of arches, is the main structural component. See Fig. 26.18. The structural principles associated with arches have been discussed above. The arches are often made of masonry (brick or stone) but can be made of steel or concrete. A distinction has to be drawn between bridges where the bridge deck is above the arch (in which case the horizontal thrusts at the supports need to be resisted by the abutments) and bridges where the bridge deck is below the arch (where the horizontal thrusts can be accommodated by tension in the bridge deck, as in Fig. 26.6).

Fig. 26.19   **Ribblehead Viaduct, North Yorkshire – a classic series of brick arches.**

Fig. 26.20   **Clifton suspension bridge, Bristol.**

Where a wide valley has to be crossed at high level a viaduct comprising a series of arches is constructed, as shown in Fig. 26.19.

Suspension bridges are essential for the longest spans. These comprise two supporting towers (of steel or concrete) over which pass steel cables. The bridge deck (steel or concrete) is suspended from these cables. The steel cables are in tension and support the huge load from the entire bridge. This load has to be sustained by massive anchorage blocks at the two ends of the bridge. Bristol's Clifton Suspension Bridge (Fig. 26.20) is an early example, and a rudimentary suspension bridge in a rural setting is shown in Fig. 26.21.

Fig. 26.21 **Suspension footbridge near Addingham, West Yorkshire.**

Fig. 26.22 **Cable stayed footbridge, Melbourne Airport, Australia.**

Cable stayed bridges look similar to suspension bridges. The bridge deck is supported by cables in tension, which are themselves supported by bridge towers. A simple cable stayed bridge is shown in Fig. 26.22, and the more complex Gateshead Millennium Bridge is shown in Fig. 26.23.

Fig. 26.23 Gateshead Millennium Bridge.
A curved-in-plan pedestrian walkway deck is suspended by cables from a high steel arch; to allow the passage of shipping the whole structure rotates through a 45° vertical angle, thus mimicking the opening and closing of a human eyelid.

Fig. 26.24 Tower Bridge, London.
An early example of a bascule bridge, in which the two halves of the main deck open by swinging upwards; huge counterweights aid this process.

Parts of some bridges have to be moveable to allow the passage of boats and ships through them. A well-known example is London's Tower Bridge, shown in Fig. 26.24.

See Figure 26.25 for illustrations of these bridge types.

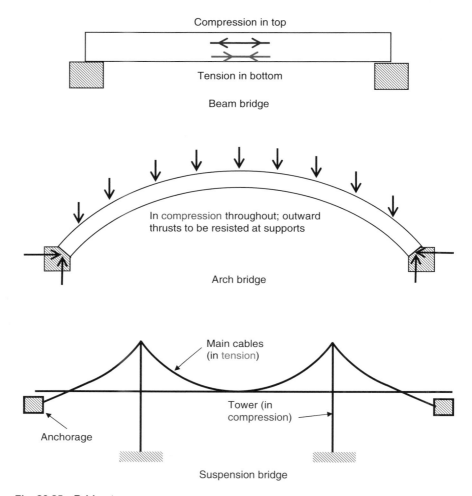

Compression in top

Tension in bottom

Beam bridge

In compression throughout; outward
thrusts to be resisted at supports

Arch bridge

Main cables
(in tension)

Tower (in
compression)

Anchorage

Suspension bridge

Fig. 26.25   Bridge types

Dams

A dam is a barrier built across a valley to form an artificial lake as a reservoir of water.
As well as water storage, dams are often used for the production of hydro-electric
power.

There are several types of dam. **Embankment dams** comprise a large dyke of earth
and rock with some form of waterproof membrane and, like **concrete gravity dams**,
rely on their own dead weight to resist the overturning moment caused by the massive pressure of the retained water.

The name **arch dam** can be confusing until it is realised that the arch is in plan.
The dam acts as an arch spanning horizontally between the valley sides, which resist
the horizontal thrusts from the water pressure. An arch dam may be single- or

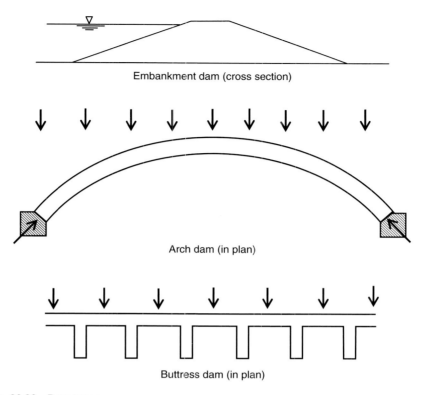

Embankment dam (cross section)

Arch dam (in plan)

Buttress dam (in plan)

Fig. 26.26 Dam types

double-curvature, the latter meaning that as well as curving in plan the dam may also curve in section. Arch dams are suited to narrow valleys with solid rock walls.

In *buttress dams* the main strength comes from vertical stiffening elements in the form of large concrete or masonry buttresses at intervals along the length of the dam, with reinforced concrete slabs spanning between these buttresses. A variation on this is the buttress arch dam, which involves horizontal arches spanning between the buttresses. See Figure 26.26 for illustrations of all these dam types.

Tunnels

There are several different types of tunnels, usually identified by the tunnelling technique used. *Bored* tunnels are usually the only option for deep tunnels, and are drilled with a machine which progressively cuts through the rock. Tunnels are lined with precast concrete segments which, when installed, comprise a series of rings which are in compression.

Shallow tunnels which do not pass under water can be constructed by the *cut and cover* technique, which is exactly what the name suggests. A deep, wide trench is dug, the tunnel is constructed inside the trench (usually in-situ concrete, rectangular in cross section), and earth is backfilled around and above the completed tunnel.

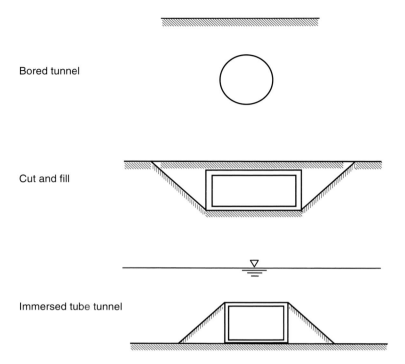

Bored tunnel

Cut and fill

Immersed tube tunnel

Fig. 26.27 Tunnel types.

Immersed tube tunnels are laid on the bed of, typically, a river estuary. Such tunnels are constructed elsewhere, usually in a nearby dry dock, in segments. When complete the segments are floated and towed to the site of the tunnel then allowed to sink into position. Examples of tunnels constructed by this method include the new Tyne Tunnel in north-east England.

See Figure 26.27 for illustrations of all these tunnel types.

27 An introduction to deflection

The term **deflection** has been mentioned several times already in this book. If load is applied to a beam or slab it will bend, and therefore deflect. The deflection is the vertical distance that the beam or slab has moved away from its originally horizontal position (as shown in Fig. 27.1) and is usually measured in millimetres.

As has been suggested previously, it is essential that deflections are minimised. Excessive deflections of floors can be alarming, and excessive deflections of door lintels can lead to distortion of door frames, making doors difficult to open and close properly. In the design of reinforced concrete floor slabs, deflection is often the critical criterion.

It is therefore important to be able to calculate deflection. The calculation of deflections normally features in Level 2 of civil engineering HND and degree courses, and therefore is beyond the scope of this book. However, I was recently struck by the difficulty that an otherwise competent group of second year students was experiencing in understanding deflection, hence this chapter.

There are several different techniques for calculating deflections. The one I'm going to teach you in this chapter is Macaulay's method, as it is both the easiest to understand and the most universally applicable. Before we get into it though you need to ensure that you understand basic calculus. If you do, feel free to skip the next section.

Basic calculus: a reminder

In your studies of mathematics you will have come across two forms of calculus: **differentiation** and **integration**. One is the reverse of the other.

Differentiation relates to **rate of change**, and the symbol d is used to represent 'change in'. Velocity (v), or speed, for example, is change in distance moved over a given change in time, and could be expressed as

$$\frac{d(\text{distance})}{dt}$$

where t represents time. This expression is known as the **differential coefficient** or the **derivative**. This relationship between distance and time is reflected in the units of velocity: miles per hour, or metres per second.

Basic Structures, Second Edition. Philip Garrison.
© 2011 John Wiley & Sons, Ltd. Published 2011 by John Wiley & Sons, Ltd.

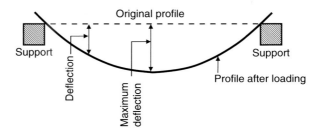

Fig. 27.1   Deflection of a beam.

Acceleration relates to the rate of change of velocity with respect to time. If v represents velocity then acceleration could be expressed as

$$\frac{dv}{dt}$$

Acceleration then is change in the change in distance with respect to time, which could be expressed as

$$\frac{d^2(\text{distance})}{dt^2}$$

which is the *second derivative* of distance with respect to time.

Where x is an unknown variable, and y is a function of $x - f(x)$ in mathematical terminology – the differential coefficient, or derivative, is represented as $\frac{dy}{dx}$. The general rule is:

If $y = x^n$, then $\frac{dy}{dx} = nx^{n-1}$

So if $y = x^3$, then $\frac{dy}{dx} = 3x^2$

Any constants (or plain numbers) disappear under the differentiation process because they are not changing.

So if $y = 2x^5 + 47$, then $\frac{dy}{dx} = 10x^4$; note that the 47, being a constant, has disappeared.

Integration is the reverse of differentiation, and is used to find areas under curves, etc. The general rule with integration is: *increase the power by one, and divide by the new power*:

$$\int x^n.dx = \frac{x^{n+1}}{n+1} + C$$

where dx means 'with respect to x' and C is a constant (i.e. a number) generated by the integration process.

A couple of further examples:

$$\int x^6.dx = \frac{x^7}{7} + C$$

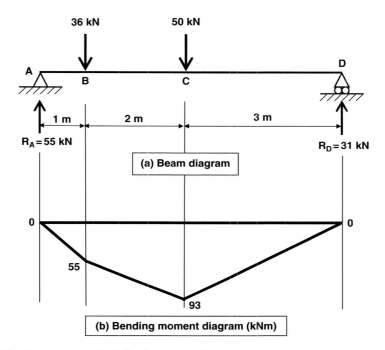

Fig. 27.2   **Example used in deflection calculation.**

$$\int 10x^4 .dx = \frac{10x^5}{5} + C = 2x^5 + C$$

Equations for bending moment

Now we're going to use our knowledge of how to calculate bending moments. If you need to revise this, please revisit Chapter 16.

Consider the beam shown in Fig. 27.2, which spans between end supports at A and D, and supports point loads of 36 kN and 50 kN at points B and C. You should be able to calculate the reactions (if not, read Chapter 9 again), and you'll find that the reaction at A (R_A) is 55 kN and the reaction at D (R_D) is 31 kN.

You can now calculate the bending moments at points A, B, C and D. The moments at A and D are zero. The moments at B and C are calculated as follows:

$$M_B = (55\,kN \times 1\,m) = 55\,kN.m$$

$$M_C = (55\,kN \times 3\,m) - (36\,kN \times 2\,m) = 93\,kN.m$$

The bending moment diagram is shown in Fig. 27.2. If you in any doubt about these calculations, see Chapter 16.

Now we are going to try to derive an equation for the bending moment at any point along the length of the beam. We'll express the bending moment in terms of x, where x is the distance (in metres) along the beam from its left-hand end. Going back to first

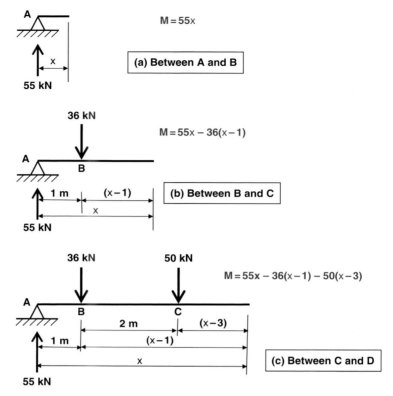

Fig. 27.3 Bending moments in different parts of a beam.

principles, remember that the moment at any point is the sum of the clockwise moments to the left of the point minus the sum of the anticlockwise moments to the left of the point.

Between A and B, $M = 55x$ (see Fig. 27.3a).

Between B and C, $M = 55x - 36(x - 1)$ (see Fig. 27.3b).

Between C and D, $M = 55x - 36(x - 1) - 50(x - 3)$ (see Fig. 27.3c).

So we've generated three separate equations for the bending moment, depending on which part of the beam we wish to find the moment in. This is a bit messy. It would be much neater if we could derive just one equation which would produce the moment at any point along the entire beam.

Of course, we are not the first people to encounter this problem. A couple of centuries ago a man named Macaulay studied it for a while and came up with a solution which is devastating in its simplicity. It centres on the use of what we now call *Macaulay brackets*, which have a specific meaning.

Notations for Macaulay brackets differ. When I was a student (many years ago) curly brackets {} were used, and some textbooks use angle brackets <>, but square brackets [] tend to be used now. Where you see square brackets in this context, remember that they have a specific meaning

The specific meaning of Macaulay brackets [] is as follows:

- If the figure in the Macaulay brackets is negative, ignore it.
- If the figure in the Macaulay brackets is zero or positive, include it.

Using Macaulay brackets, the equation for the bending moment at any point in the above beam is:

$$M = 55x - 36[x - 1] - 50[x - 3]$$

Check this equation for $x = 0$, $x = 1$, $x = 3$ and $x = 6$ to ensure that it gives the same values as in the bending moment diagram above (Fig. 27.2).

Deflection in beams

So, what have we covered so far in this chapter? Two things, essentially. We've reviewed our knowledge of calculus and we've derived an equation for the bending moment at any point along a beam – with a little help from Mr Macaulay. Where is this all leading? It leads to deflection in beams.

It can be shown that:

$$M = - EI\frac{d^2v}{dx^2}$$

where:

M = bending moment
E = Young's modulus
I = second moment of area
x = distance along beam measured from the left-hand end
v = deflection of beam at point x.

The equation can be rearranged to read:

$$\frac{d^2v}{dx^2} = -\frac{M}{EI}$$

This equation is the key to calculating deflections using Macaulay's method. In the above example we derived an equation for the moment at any point along a beam in terms of x. We could substitute it for M in the above equation to get an equation for $\frac{d^2v}{dx^2}$ in terms of x and EI. Integrating the equation once would give us an expression for $\frac{dv}{dx}$, and integrating a second time gives us an expression for the deflection v in terms of x and EI. So, if we know E and I, we can calculate the deflection, in millimetres, at any given point along the beam.

Let's apply this to the above example.

We saw earlier that the moment M at any point along the beam shown in Fig. 27.2 is:

$$M = 55x - 36[x - 1] - 50[x - 3]$$

Substituting this in the equation

$$\frac{d^2v}{dx^2} = -\frac{M}{EI}$$

gives:

$$\frac{d^2v}{dx^2} = -\frac{1}{EI}\{55x - 36[x-1] - 50[x-3]\}$$

Now integrate this expression, treating the terms within the squared brackets as a single term (don't expand the term in brackets):

$$\frac{dv}{dx} = -\frac{1}{EI}\left\{\frac{55x^2}{2} - \frac{36[x-1]^2}{2} - \frac{50[x-3]^2}{2} + A\right\}$$

where A is a constant generated by the integration process. We shall determine the value of A later.

Now integrate the expression one more time:

$$v = -\frac{1}{EI}\left\{\frac{55x^3}{6} - \frac{36[x-1]^3}{6} - \frac{50[x-3]^3}{6} + Ax + B\right\}$$

where B is a further constant of integration. Simplifying the above expression gives:

$$v = -\frac{1}{EI}\{9.167x^3 - 6[x-1]^3 - 8.33[x-3]^3 + Ax + B\}$$

This then is an expression for the deflection (v) at any point x along the length of the beam. However, we don't yet know the values of A and B, and the equation isn't usable to us until we find these out. Here's how.

Do we already know the deflection at any particular points along the beam? Yes, we do. At the two supports A and D (where $x = 0$ and $x = 6$ m respectively) we know that the deflection is zero. This is because the beam cannot move downwards at a support. These known values are called the **boundary conditions**.

We can substitute these values in the above equation, as follows:

At point A, when $x = 0$, $v = 0$:

$$0 = -\frac{1}{EI}\{0 - 0 - 0 + 0 + B\}$$

Therefore $B = 0$.

At point B, when $x = 6$, $v = 0$:

$$0 = -\frac{1}{EI}\{(9.167 \times 6^3) - (6 \times 5^3) - (8.33 \times 3^3) + 6A\}$$

$$0 = -\frac{1}{EI}\left\{1980 - 750 - 224.9 + 6A\right\}$$

As the terms in the curly brackets must add up to zero, $A = -167.5$.
Therefore the deflection equation for this problem is now:

$$v = -\frac{1}{EI}\left\{9.167x^3 - 6[x-1]^3 - 8.33[x-3]^3 - 167.5x\right\}$$

We can use this expression to calculate the deflection at any point in the beam in terms of EI. As the actual deflection in any beam is dependent on the material it is made of (represented by Young's modulus E) and its sectional geometry (represented by second moment of area I), then we can determine the deflection in millimetres only if we know the values of E and I.

Determination of maximum deflection

Normally we are interested in determining the *maximum* deflection in any beam. Before we can do this, we have to know *where* the maximum deflection occurs. If the beam is simply supported and the loading is symmetrical then the maximum deflection will occur at the midpoint of the beam. If the loading is not symmetrical (as in this case), then we first of all have to determine the *position* of maximum deflection before we can calculate its value.

(Note: the position of maximum deflection is not necessarily the same as the position of maximum bending moment.)

The term $\frac{dv}{dx}$ stands for the 'change in deflection (i.e. vertical distance) divided by horizontal distance'; in other words, it represents the *gradient* of the deflected form at any point. (As an analogy, consider a 1 in 4 hill on a road, which means that the road rises – or falls – one metre for every four metres horizontally.) At the position of maximum deflection the slope of a tangent to the beam's deflected form is horizontal, therefore it has no gradient and $\frac{dv}{dx}$ is zero. To identify this point, we have to find the value of x which causes the $\frac{dv}{dx}$ equation above to have a value of zero. A simplification of that equation is:

$$\frac{dv}{dx} = -\frac{1}{EI}\left\{27.5x^2 - 18[x-1]^2 - 25[x-3]^2 - 167.5\right\}$$

If $\frac{dv}{dx} = 0$, then $27.5x^2 - 18[x-1]^2 - 25[x-3]^2 - 167.5 = 0$

In this case, it is possible to solve the above equation by multiplying out the brackets, reducing it to a quadratic equation, and solving it using the standard formula. In some cases it's not easy to do this, so an alternative approach (particularly if you didn't understand the last sentence!) would be to try an iterative approach. In other words, insert certain values for x in the above equation and calculate the result.

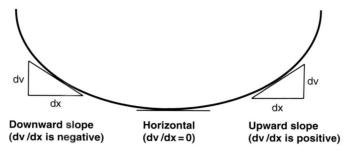

Fig. 27.4   Slope of a deflected beam.

When $x = 1$, $\dfrac{dv}{dx} = -140$

When $x = 3$, $\dfrac{dv}{dx} = +8$

As these two results have different signs (that is, one is negative and the other is positive) they represent downward and upward sloping parts of the beam respectively. See Fig. 27.4. So the position of maximum deflection lies somewhere between these two points, and evidently quite close to $x = 3$ m.

When $x = 2.95$, $\dfrac{dv}{dx} = +3.4$

When $x = 2.90$, $\dfrac{dv}{dx} = -1.2$

So the point of maximum deflection lies somewhere between $x = 2.90$ and 2.95, and further investigation will identify the point of maximum deflection as $x = 2.91$ m. We can substitute this value into the deflection equation above to get the value of maximum deflection:

$$v = -\frac{1}{EI}\left\{9.167x^3 - 6[x-1]^3 - 8.33[x-3]^3 - 167.5x\right\}$$

$$v = -\frac{1}{EI}\left\{(9.167 \times 2.91^3) - (6 \times 1.91^3) - 0 - (167.5 \times 2.91)\right\}$$

which simplifies down to:

$$v = \frac{303}{EI}$$

Now if a steel beam of designation $305 \times 165 \times 40$ is used (subject to checking that it can sustain the bending moments and shear forces applied to it), section property tables produced by Tata show that the second moment of area value for that standard section is $I = 8503\ \text{cm}^4 = 8503 \times 10^{-8}\ \text{m}^4$. If the Young's modulus value of steel is $E = 210\,\text{kN/mm}^2$, which is $210 \times 10^6\,\text{kN/m}^2$, then the value of EI can be calculated as follows:

$EI = 210 \times 10^6 \times 8503 \times 10^{-8}$

$EI = 17856.3\,\text{kN.m}^2.$

(EI should always be expressed in units of kN.m^2 for consistency.)
So the actual maximum deflection, v, is calculated as follows:

$$v = \frac{303}{EI} = \frac{303}{17856.3} = 0.0169\,\text{m} = 16.9\,\text{mm}$$

Codes of practice currently decree that deflection should not exceed span/350. So in this case the maximum permissible deflection $= 6000/350 = 17.1\,\text{mm}$. Our maximum deflection of 16.9 mm is less than this and is therefore allowable.

Summary of the use of Macaulay's method

1) Derive an equation for the bending moment (M) at any point along the beam in terms of x.
2) Substitute this equation into $\dfrac{\text{d}^2v}{\text{d}x^2} = -\dfrac{M}{EI}$
3) Integrate twice to obtain an expression for deflection v.
4) Substitute boundary conditions to find constants A and B.
5) Find *position* of maximum deflection by making the $\dfrac{\text{d}v}{\text{d}x}$ equation equal to zero, then solving for x.
6) Find the *value* of the maximum deflection by substituting this value of x in the deflection (v) equation.
7) If E and I are known, use their values to calculate the actual maximum deflection of the beam in millimetres.

Maximum deflection for standard cases

We will now use Macaulay's method to calculate the maximum deflection for the following two cases:

1) beam with central point load
2) beam with uniformly distributed load over its entire length

Due to symmetry we know that the maximum deflection for each of these two cases will occur half way along the beam.

Case 1: Central point load

A simply supported beam of length L is subjected to a central point load of P, as shown in Fig. 27.5a. The bending moment diagram is shown in Fig. 27.5b; see Chapter 16 if you need to remind yourself how this was obtained.

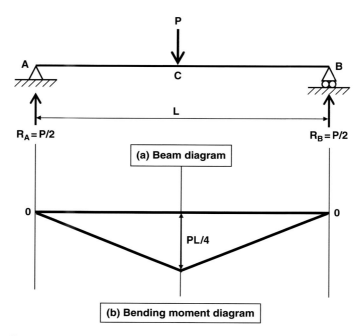

Fig. 27.5   **Bending moment diagram for a beam carrying a central point load.**

Using Macaulay notation, the moment at any point x is:

$$M = \frac{Px}{2} - P\left[x - \frac{L}{2}\right]$$

Since $\dfrac{d^2v}{dx^2} = -\dfrac{M}{EI}$:

$$\frac{d^2v}{dx^2} = -\frac{1}{EI}\left\{\frac{Px}{2} - P\left[x - \frac{L}{2}\right]\right\}$$

$$\frac{dv}{dx} = -\frac{1}{EI}\left\{\frac{Px^2}{4} - \frac{P}{2}\left[x - \frac{L}{2}\right]^2 + A\right\}$$

$$v = -\frac{1}{EI}\left\{\frac{Px^3}{12} - \frac{P}{6}\left[x - \frac{L}{2}\right]^3 + Ax + B\right\}$$

Boundary conditions: when $x = 0$, $v = 0$. Substitution in above equation gives $B = 0$.

When $x = L$, $v = 0$. Substitution in the above equation gives $A = -\dfrac{PL^2}{16}$
The deflection equation now becomes:

$$v = -\frac{1}{EI}\left\{\frac{Px^3}{12} - \frac{P}{6}\left[x - \frac{L}{2}\right]^3 - \frac{PL^2 x}{16}\right\}$$

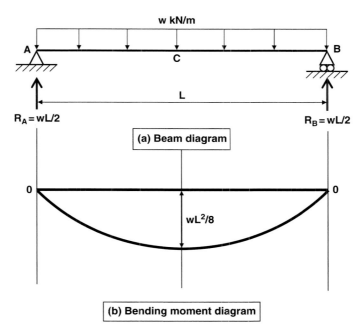

Fig. 27.6   **Bending moment diagram for a beam carrying a uniformly distributed load.**

By inspection, maximum deflection occurs at midspan, i.e. at $x = L/2$. Substituting this value into the above equation gives:

Maximum deflection $= \dfrac{PL^3}{48EI}$.

Case 2: UDL over entire length of beam

A simply supported beam of length L is subjected to a uniformly distributed load of w per metre, as shown in Fig. 27.6a. The bending moment diagram is shown in Fig. 27.6b; again, see Chapter 16 if you need to remind yourself how this was obtained.

Using Macaulay notation, the moment at any point x is:

$$M = \frac{wLx}{2} - \frac{wx^2}{2}$$

Since $\dfrac{d^2v}{dx^2} = -\dfrac{M}{EI}$:

$$\frac{d^2v}{dx^2} = -\frac{1}{EI}\left\{\frac{wLx}{2} - \frac{wx^2}{2}\right\}$$

$$\frac{dv}{dx} = -\frac{w}{2EI}\left\{\frac{Lx^2}{2} - \frac{x^3}{3} + A\right\}$$

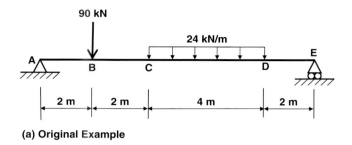

(a) Original Example

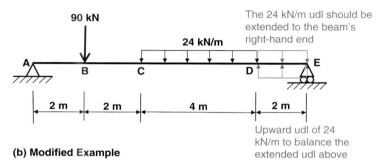

(b) Modified Example

Fig. 27.7   Deflection example with uniformly distributed load.

$$v = -\frac{w}{2EI}\left\{\frac{Lx^3}{6} - \frac{x^4}{12} + Ax + B\right\}$$

Boundary conditions: when $x = 0$, $v = 0$. Substitution in above equation gives $B = 0$.

When $x = L$, $v = 0$. Substitution in the above equation gives $A = -\dfrac{L^3}{12}$

The deflection equation now becomes:

$$v = -\frac{w}{2EI}\left\{\frac{Lx^3}{6} - \frac{x^4}{12} - \frac{L^3x}{12}\right\}$$

By inspection, maximum deflection occurs at midspan, i.e. at $x = L/2$. Substituting this value into the above equation gives:

$$\text{Maximum deflection} = \frac{5wL^4}{384EI}$$

Numerical example involving a UDL

Find the position and value of the maximum deflection of the beam shown in Fig. 27.7a.

Total load on the beam = 90 kN + (24 kN/m × 4 m) = 186 kN.

The end reactions are:

$R_A = 110.4\,\mathrm{kN}$

$R_E = 75.6\,\mathrm{kN}.$

(Revisit Chapter 9 to learn how to calculate reactions.)

We now need to produce an equation for the bending moment M at any point along the beam, but you'll find that you won't be able to do this for the problem as it stands. (Try it and you'll find out why.) In order to get round this difficulty, the 'trick of the trade' is to extend the UDL up to the right-hand end of the beam, then counteract this load by putting an equal upwards load on the extended length, as shown in Fig. 27.7b.

The moment at distance x along the beam can be determined from the following equation:

$$M = 110.4x - 90[x-2] - \frac{24}{2}[x-4]^2 + \frac{24}{2}[x-8]^2$$

Substituting this in the equation

$$\frac{d^2v}{dx^2} = -\frac{M}{EI}$$

and integrating, produces:

$$\frac{dv}{dx} = -\frac{1}{EI}\left\{\frac{110.4x^2}{2} - \frac{90}{2}[x-2]^2 - \frac{12}{3}[x-4]^3 + \frac{12}{3}[x-8]^3 + A\right\}$$

Integrating a second time produces:

$$v = -\frac{1}{EI}\left\{\frac{110.4x^3}{6} - \frac{90}{6}[x-2]^3 - [x-4]^4 + [x-8]^4 + Ax + B\right\}$$

The beam is supported at A (where $x = 0$) and E (where $x = 10\,\mathrm{m}$), so $v = 0$ at each of these points (as there can be no deflection at a support).

Substituting $v = 0$ and $x = 0$ in the above equation: $B = 0$.

Substituting $v = 0$ and $x = 10$ in the above equation: $A = -944$.

These A and B values can be substituted in the above equations.

Now we need to identify the point where the maximum deflection occurs. This occurs when $\frac{dv}{dx} = 0$, so substitute into the $\frac{dv}{dx}$ equation above to find the corresponding x value.

$$55.2x^2 - 45[x-2]^2 - 4[x-4]^3 + 4[x-8]^3 - 944 = 0$$

Unless you're a hardened mathematician the easiest way to solve this is by the iterative method outlined in the earlier example; that is, try various sensibly chosen values of x to see what $\frac{dv}{dx}$ value they give.

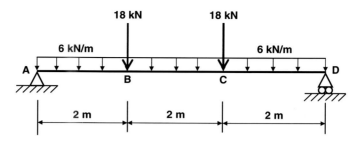

Fig. 27.8 **Tutorial example 1.**

When $x = 4$, $\dfrac{dv}{dx} = -240.8$

When $x = 5$, $\dfrac{dv}{dx} = +27$

When $x = 4.90$, $\dfrac{dv}{dx} = +0.034$

Therefore the maximum deflection occurs when $x = 4.9\,\mathrm{m}$.
  Substituting this into the deflection equation gives the maximum deflection:

$$v = -\frac{2827}{EI}$$

If the beam is made of a grade of steel for which Young's modulus $(E) = 200\,\mathrm{kN/mm^2}$, and its second moment of area $(I) = 20{,}000\,\mathrm{cm^4}$, then:

$$EI = (200 \times 10^6) \times (20\,000 \times 10^{-8}) = 40\,000\,\mathrm{kN.m^2}$$

$$v = \frac{2827}{40000} = 0.071\,\mathrm{m} = 71\,\mathrm{mm}$$

Tutorial examples

1) Use Macaulay's method to calculate the maximum vertical deflection of the beam shown in Fig. 27.8 under the loading indicated. Express your answer in terms of EI. (Hint: note the symmetry of the beam and its loading.)
2) Use Macaulay's method to calculate the maximum vertical deflection of a cantilever of length L subjected to a point load P at its free end.

Tutorial answers

1) $v = 239.25/EI$: integration constants: $A = -126$, $B = 0$.
2) $v = PL^3/3EI$: integration constants: $A = B = 0$.

Further reading

The following listing is not designed to be exhaustive; it simply gives some titles that the interested reader might find useful in moving on (in various directions) from this book.

Hulse R & Cain J: *Structural Mechanics* (Palgrave, Second Edition 2000). This title moves the *Basic Structures* reader on to material taught in Level 2 (and some Level 3) on university civil engineering degree courses.

The following five titles relate to structural behaviour as a physical concept, with little or nothing in the way of mathematical calculations. The first two will be of particular interest to students of architecture.

- Silver P, McLean W: *Introduction to Architectural Technology* (Laurence King 2008)
- Charleson A: *Structure as Architecture* (Elsevier 2005)
- Yeomans D: *How Structures Work* (Wiley-Blackwell 2009)
- Millais M: *Building Structures* (Spon, Second Edition 2005)
- Gordon J: *Structures* (Penguin 1978)

Draycott T & Bullman P: *Structural Elements Design Manual* (Elsevier 2009). In my opinion this is the most concise and student-friendly book on structural design, which is normally studied at Level 2 of civil engineering degree courses and is outlined in Chapter 25 of *Basic Structures*.

The following two titles are American and thus will be of interest to readers whose course of study follows the American pattern. Whilst the underlying concepts are universal, the approach, symbols, standards and units of measurement are different in the USA from those familiar to British engineers. The first title deals with structural analysis, the second covers structural design.

- Schodek D: *Structures* (Prentice Hall, Fourth Edition 2001)
- Ochshorn J: *Structural Elements for Architects and Builders* (Elsevier 2010)

Blockley D: *Bridges* (Oxford University Press 2010) is a very descriptive book on how bridge structures work, with plenty of examples.

Basic Structures, Second Edition. Philip Garrison.
© 2011 John Wiley & Sons, Ltd. Published 2011 by John Wiley & Sons, Ltd.

The following two titles, which are descriptive and based on examples, tell us why (or perhaps how) buildings stand up, and why they (very occasionally) fall down.

- Salvadori M: *Why Buildings Stand Up* (Norton 1980)
- Salvadori M: *Why Buildings Fall Down* (Norton 1992)

Appendix 1   Weights of common building materials

A full list is given in British Standard BS 648. The more commonly used materials are discussed below. (Please note that these figures are 'typical' only, as the strength of any material varies according to the type or grade of material.)

Reinforced concrete

Unit weight: $24\,kN/m^3$ ($2400\,kg/m^3$)
    Therefore a 100 mm thick concrete wall weighs $2.4\,kN/m^2$ ($240\,kg/m^2$).

Blockwork

Unit weight: $22\,kN/m^3$ ($2200\,kg/m^3$)
    Therefore a 100 mm thick blockwork wall weighs $2.2\,kN/m^2$ ($220\,kg/m^2$).
    Lightweight (aerated) blockwork can weigh considerably less (as low as $6\,kN/m^3$).

Brickwork

Approximately the same weight as blockwork (see above).

Steel

Unit weight: $78.5\,kN/m^3$ ($7850\,kg/m^3$)
    Steel beams weigh between 0.2 and $2.0\,kN/m$ (20–200 kg/m) depending on size.

Aluminium

Unit weight: $27.7\,kN/m^3$ ($2771\,kg/m^3$)

Basic Structures, Second Edition. Philip Garrison.
© 2011 John Wiley & Sons, Ltd. Published 2011 by John Wiley & Sons, Ltd.

Timber

Unit weight: Softwood: 5.9 kN/m^3 (590 kg/m^3). Hardwood: 12.5 kN/m^3 (1250 kg/m^3)

Therefore a 50 mm × 200 mm ('two by eight') softwood joist weighs 0.06 kN/m (6 kg/m).

Glass

Unit weight: 25 kN/m^3 (2500 kg/m^3)

Therefore the weight of glass is 0.025 kN (2.5 kg) per millimetre thick.

Water

Unit weight: 10 kN/m^3 (1000 kg/m^3)

Live loads

Live loads (i.e. non-permanent loads due to people and furniture in a room in a building) are assumed to be uniformly distributed and are expressed in kN/m^2. Values of live load depend on the use of the building (or part of the building) concerned. A full listing appears in British Standard BS 6399 Part 1. Some values are given below.

- domestic: 1.5 kN/m^2
- offices: 2.5 kN/m^2
- cafes/restaurants: 2.0 kN/m^2
- classrooms: 3.0 kN/m^2
- assembly: fixed seating: 4.0 kN/m^2
- corridors/stairs in hotels, etc.: 4.0 kN/m^2
- exhibitions: 4.0 kN/m^2
- gymns: 5.0 kN/m^2
- bars, concert halls, etc.: 5.0 kN/m^2
- stages: 7.5 kN/m^2
- shops: 4.0 kN/m^2
- parking (cars): 2.5 kN/m^2
- plant rooms: 7.5 kN/m^2.

Note: British Standards can be viewed on the internet at *www.athens.ac.uk*. To access this site, an 'Athens password' is required, which can be obtained by students and staff at UK universities. See your university learning centre for details.

Appendix 2　Conversions and relationships between units

Inches, feet and metres

1 inch = 25.4 mm
1 foot = 304.8 mm = 0.3048 metres
1 metre = 3.281 feet
$1\,m^2 = 10.76\,ft^2$
$1\,ft^2 = 0.092\,m^2$

Yards and metres

1 yard = 3 feet = 36 inches = 0.9144 metres
1 metre = 1.094 yards
$1\,yd^2 = 0.836\,m^2$
$1\,m^2 = 1.196\,yd^2$

Acres and hectares

$1\,acre = 4840\,yd^2 = 4047\,m^2$
$1\,hectare = 10{,}000\,m^2 = 2.47\,acres$
1 acre = 0.405 hectares

Miles and kilometres

1 mile = 1760 yards = 1609.3 metres
1 km = 1000 metres
1 mile = 1.6093 km
1 km = 0.621 miles

Litres and cubic metres

1 metre = 100 cm
$1\,m^3 = 10^6\,cm^3$

Basic Structures, Second Edition. Philip Garrison.
© 2011 John Wiley & Sons, Ltd. Published 2011 by John Wiley & Sons, Ltd.

$$1 \text{ millilitre} = 1 \text{ cm}^3$$
$$1 \text{ litre} = 1000 \text{ millilitres} = 1000 \text{ cm}^3$$
$$1000 \text{ litres} = 1 \text{ m}^3$$

Pounds, kilograms and stones

$$1 \text{ lb} = 0.454 \text{ kg} = 454 \text{ g}$$
$$1 \text{ kg} = 2.203 \text{ lbs}$$
$$1 \text{ stone} = 14 \text{ lb} = 6.356 \text{ kg}$$

Kilograms, kN and tonnes

$$10 \text{ N} = 1 \text{ kg}$$
$$1000 \text{ N} = 1 \text{ kN}$$
$$10 \text{ kN} = 1 \text{ tonne} = 1000 \text{ kg}$$

Tons and tonnes

$$1 \text{ ton} = 160 \text{ stone} = 1017 \text{ kg}$$
$$1 \text{ tonne} = 0.983 \text{ tons}$$
$$1 \text{ ton} = 1.017 \text{ tonnes}$$

Appendix 3 Mathematics associated with right-angled triangles

Pythagoras' theorem

This states that 'the square of the hypotenuse of a right-angled triangle is equal to the sum of the squares of the other two sides'. In plain English this means that if the length of any two sides of a right-angled triangle are known, then the length of the third side can be determined using the relationships shown in Fig. A1.

Basic trigonometry

With respect to a right-angled triangle, sines, cosines and tangents (normally abbreviated to sin, cos and tan respectively) are defined in Fig. A2. There is nothing 'magic' about this. A sine, cosine or tangent is simply the ratio of the lengths of two sides of a right-angled triangle.

Suppose that we are interested in the angle formed between two sides of the triangle. The angle is represented by the Greek letter θ and is measured in degrees. 'Hypotenuse' represents the length of the longest side of the right-angled triangle – which is always the side opposite to the right angle. 'Opposite' represents the length of the side opposite the angle θ. 'Adjacent' represents the length of the side adjacent to the angle θ.

For example, if the 'opposite' side is 2 metres long and the 'hypotenuse' is 2.5 metres long, then $\sin \theta = 2.0/2.5 = 0.8$.

The reader should refer to a basic mathematics textbook if further information is required.

Basic Structures, Second Edition. Philip Garrison.
© 2011 John Wiley & Sons, Ltd. Published 2011 by John Wiley & Sons, Ltd.

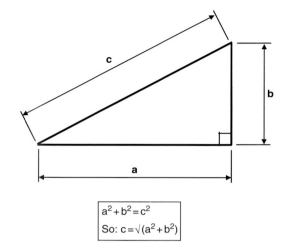

$$a^2 + b^2 = c^2$$
$$\text{So: } c = \sqrt{(a^2 + b^2)}$$

Fig. A1   Pythagoras' theorem.

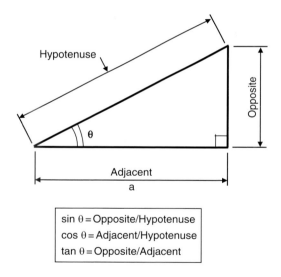

$$\sin\theta = \text{Opposite/Hypotenuse}$$
$$\cos\theta = \text{Adjacent/Hypotenuse}$$
$$\tan\theta = \text{Opposite/Adjacent}$$

Fig. A2   Sines, cosines and tangents.

Appendix 4   Symbols

The units normally used in structural mechanics are given in brackets after each definition.

A = cross-sectional area (mm^2)
E = Young's modulus or modulus of elasticity (kN/mm^2)
I = second moment of area (mm^4)
L = length; span of beam or slab (millimetres or metres)
M = moment (kN.m)
P = force (kN)
R = reaction (kN)
V = shear force (kN)
w = uniformly distributed load per metre (kN/m)
W = total uniformly distributed load per metre (kN)
σ = stress (direct or bending) (N/mm^2)
τ = shear stress (N/mm^2)

Basic Structures, Second Edition. Philip Garrison.
© 2011 John Wiley & Sons, Ltd. Published 2011 by John Wiley & Sons, Ltd.

Appendix 5   A checklist for architects

The hints below don't apply exclusively to architects of course. The reason for the title is that architectural students tend to get involved in large conceptual projects. The following checklist might be useful in any project work you are doing.

- Does your building have a *structure*? The structure might be overt or concealed, but it must exist, or the building will not stand up.
- Is it a *framed structure*, and, if so, is this frame of steel or concrete (or other material)? What are the reasons for your choice?
- Does your project include *cantilevers* of significant length? How are these supported? Are they justified both in terms of the use of the building and economically?
- Does your project contain an atrium or other elements including significant areas of *glazing*? How is this glazing supported? Again, is this justified by the use of the building and economically?
- What *foundation* type is being used?
- Are there any *long span*s (>10 metres) in your building? What is the form of the beams or slabs spanning this distance, and how are they, in turn, supported?
- What are the indicative *sizes* of the various beams, columns, slabs and walls in your building?
- What is the form of your *roof*? Is it flat or sloping? What material is the roof made out of, and how is it supported?
- What *flooring systems* are you adopting?
- Indicate the *load paths* within your structure.
- Ensure that all parts of your structure are adequately *supported*.
- How is *stability* of your structure achieved?

Basic Structures, Second Edition. Philip Garrison.
© 2011 John Wiley & Sons, Ltd. Published 2011 by John Wiley & Sons, Ltd.

Index